One Origin of Digital Humanities

Julianne Nyhan · Marco Passarotti
Editors

One Origin of Digital Humanities

Fr Roberto Busa in His Own Words

Foreword by Steven E. Jones

Springer

Editors
Julianne Nyhan
University College London (UCL)
London, UK

Marco Passarotti
Università Cattolica del Sacro Cuore
Milan, Italy

Translated by Philip Barras, Andreia Carvalho, and Tessa Hauswedell

ISBN 978-3-030-18311-0 ISBN 978-3-030-18313-4 (eBook)
https://doi.org/10.1007/978-3-030-18313-4

Typeset by Servis Filmsetting Ltd, Cheshire

This Springer imprint is published by the registered company Springer Nature Switzerland AG
The registered company address is: Gewerbestrasse 11, 6330 Cham, Switzerland

For Reimar, Joey, Clara, Iris, John and Eileen
and
for Nina, Ilde, Maria Assunta, Carlo and Alice

Table of Contents

List of Figures

List of Tables

Foreword

In a 1962 essay included in the present collection, Father Roberto Busa, S.J., looked back at the beginnings of his project in 1949 and admitted: "I was unaware of the fact that I was placed in the sequence of events by which the automation of accounting caused the worldwide evolution of the means of information" (see p. 80). That term, "world" or "worldwide" (*mondiale*, and elsewhere in the same essay, *tutto il mondo*), describes a technological shift, but also the pioneering scholar's own ambitions for his experiments using machinery to analyse language—what were sometimes called in English "literary (or linguistic) data processing." Those ambitions were global.

The ambitions are reflected in the Busa Archive, which Father Busa himself first organized by national culture or language. When his work with IBM began, his own English was not yet very strong. Busa himself later remarked that the English translation made by someone else for his first major research publication in 1951 (see Chap. 2), was often awkward (he called it "hilarious"), but that he couldn't tell at the time (Roberto Busa to Robert D. Eagleson, July 4, 1966). He would soon become fluent in English, as he already was in several other world languages. The languages represented in the archive include not only modern English and European tongues, but, as we might expect, Jesuit-to-Jesuit Latin, including some of his earliest correspondence in the 1940s with fellow priests in North America, paving the way for his transatlantic research program. Half a century later, Father Busa would characterize his own early work as part of the emergence of linguistic—as distinct from numerical and scientific—data processing, a "spark . . . which has developed into a blaze of activity that now covers the entire life of the world." (Busa, unpublished autobiographical manuscript).

That image of the "blaze" echoes the famous Jesuit charge attributed to St. Ignatius, to "Go forth and set the world on fire." Father Busa's global ambitions were a product of his vocation but also of his historical moment. Although he claimed to be "the first and only one in the world to venture to saddle the flying horse with lexicology," he also acknowledged that, "[i]f it did not come to me, the idea certainly would have come to someone else, and perhaps one day it may be known that it came to someone before me, to whom nobody at the time had paid any attention" (see p. 80). His true contribution to scholarship, he says, was "patience," a diligent application which allowed him over time to transform the "idea" of linguistic data processing "into a mature and practical methodology that can be applied, so to speak, to a production line" (see p. 80).

Busa arrived in New York City (by way of Canada) in the autumn of 1949, not long after regular transatlantic passenger voyages had resumed following the war. After a series of inquiries and referrals he found his way to IBM World Headquarters at 590 Madison Avenue and to the office of the company's founder, Thomas J. Watson, Sr. It was an auspicious moment for the company. A plaque mounted on that building was engraved with one of Watson's favourite mottos: "World

Peace Through World Trade." Between the installation of the plaque in 1938 and late 1949, World War II had intervened, altering the implications of "World Trade" between the U.S. and Europe. Earlier in 1949, just before Busa arrived, IBM had founded a new, dedicated subsidiary organization, IBM Word Trade Corporation, with its own headquarters downtown, near the new U.N. building. It was to that new organization and its senior engineer, Paul Tasman, that Father Busa was sent after his initial meeting with Watson. Tasman and Busa were to remain friends and collaborators for decades. Tasman visited Father Busa in Italy on multiple occasions, and Father Busa presided at his American colleague's funeral in 1988.

After the war, IBM's internationalism began to morph into what we recognize as corporate multinationalism. On a practical level, this involved finding new uses for wartime assets and re-establishing and strengthening ties with Europe that had been strained during the conflict. Or displacing old ties, as in the case of IBM's business with the German data-processing subsidiary, Dehomag, under the Nazi regime. Even a very small investment in the punched-card experiments of an ambitious Italian priest who wanted to process medieval Latin texts might have seemed to IBM like a logical result of the company's own global ambitions, a contribution, however modest, to the company's postwar strategy in Europe and the U.S. A decade after the agreement was reached, one 1960 letter from the younger son of the company's founder, Arthur K. ("Dick") Watson, to Father Busa revealed another benefit of the investment—good marketing. In the letter Watson politely refuses Father Busa's latest request for additional funding, though he promises additional machinery and time on machines in New York. He expresses respect for the "pioneering work" Busa has done and acknowledges an area of significant mutual interest: "We have always kept in mind, not only the humanistic value of this work that you are doing, but also the very favourable publicity that it provided both IBM and the Center for the Automation of Literary Analysis." (Arthur K. Watson to Roberto Busa, April 7, 1960).

Father Busa's project contributed in some measure to technical developments within IBM, including Peter Luhn's Key Word In Context (KWIC) protocol for information retrieval, and experiments in Machine Translation (MT) in the 1950s and 1960s. Indeed, data input for Machine Translation was carried on at Busa's own centre, CAAL (the Centro per L'Automazione dell'Analisi Letteraria, or Center for the Automation of Literary Analysis; later, it was sometimes translated as Linguistic Analysis). This took place by way of an arrangement Busa made that linked IBM, Georgetown University linguistics researchers, and CETIS (*Centre Européen de Traitement de l'Information Scientifique*) at the European Atomic Energy Commission (Euratom) in Ispra, Italy, established by treaty of 1957. Busa's young operators punched Russian-language texts onto cards for processing 30 kilometers away at CETIS, and in return CAAL received some funding and some operators got jobs at Euratom after leaving CAAL. This was Cold-War defence work, in addition to being scholarly research. Early humanities computing, like other forms of technology research, was deeply entangled with the emergent military-industrial complex.

It's in that context that Father Busa imagined in 1961 that CAAL might become a node in a network of linguistic data processing centres around the world. A paper published in 1962 explicitly imagines "The international services of the Centre," the first of which was "to keep each of the centres at the international level informed about the other centres and about other ongoing work worldwide" (chap 5). This plan for a networked consortium lies behind a good deal of the multilingual publications he produced in those years and it drove much of the activity of CAAL in the crucial mid-century period, from the work on some of the Dead Sea Scrolls to his public presentation in the IBM pavilion at the World's Fair in Brussels in 1958 (the first World's Fair held since the end of the war). The photograph in Figure 2 below shows Busa on the stage of that pavilion, holding a microphone and presenting his work to a large crowd. The overall theme of Expo 58 was "A World View: A New Humanism," and planning for the fair made it clear that one of its purposes was to represent Western market-driven commerce and advanced technology as more advanced *and* more "humanistic" than the alternatives in the U.S.S.R. Sputnik had been launched in the previous year. The fairgrounds were spread out around the colossal molecular-structure building known as the Atomium, with its shiny metallic spherical rooms connected by tubes (Jones, 2016, 98-100).

This was literally an international stage on which to showcase Father Busa's experiments in computing in the humanities, as opposed to the more commonly expected uses in business and the military. The pavilion included a demonstration of the IBM 305 RAMAC machine, which had multiple-disk storage and answered questions on world history in ten languages, and featured a ten-minute animated film by Eames Studios commissioned by IBM, *The Information Machine: Creative Man and the Data Processor*, which later received an award from the U.S. State Department. The film associates technology and computing with the long history—and prehistory—of human creativity. Modern society's complicated problems, including the flood of data it has to deal with, require new "tools," the film suggests, but "something has now emerged that might make even our most elegant theories workable," at which point the images cycle through abacus beads, machine cogs, vacuum tubes, and finally "the electronic calculator," a male IBM worker (in typical white shirt and tie) sitting at the console. "This is information," the voiceover says, and "the proper use of it can bring a new dignity to mankind."

Father Busa's demo at the World's Fair broadcast essentially the same message (he seems to have appeared on television that week). From the point of view of IBM the demo was clearly intended, like the Eames film, to help humanize technology at the height of the Cold War, when computing was linked in the public imagination with terrifying missiles and impersonal bureaucracies. In contrast, the colourful animated short celebrated "artists" (presaging Apple's ad campaigns decades later) whose creative thinking led to computers. Meanwhile, adjacent crowds gathered to listen to the philologist-priest talk about his experiments in literary data analysis. That same year (1958) Father Busa published a paper that he had originally given at a conference in 1956, in which he describes the humanistic use of computing: "It is the despised machine that repeats to us the invitation 'know thyself still more profoundly, scientifically and humanistically: study your speech'"—an idea which, as the editors of this volume point out, Busa "would continue to return to even in his final publications." (see p. 59).

It is not surprising that a European linguist would himself work in multiple world languages. Language was not only Father Busa's fundamental area of research; world languages were the practical means through which to construct a worldwide network of researchers and centres. In the 1950s and 1960s, while working on the Dead Sea Scrolls, he came up with the idea of distributing the necessary scholarly work of lemmatizing the Hebrew and Aramaic texts, a form of outsourcing if not quite "crowdsourcing" the linguistic work to an international community of specialists. The Busa Archive contains copies of a booklet he printed for this purpose, dated June 8, 1958, presumably for distribution to academic experts in ancient philology. "Dear Professor," it begins, followed by a formal request for collaboration in lemmatizing and sorting homographs found in the Dead Sea Scrolls texts, with instructions on how to list and return the results. This scheme for collaboration evidently failed to produce the necessary lemmatizations in time. The Dead Sea Scrolls index was never completed. But the scheme is yet another reminder of how important to Father Busa was the idea of worldwide collaboration, an idea that grew out of his sense of mission but also very much out of his historical moment—when international scientific cooperation was being put on a new footing in promising but also complicated, sometime compromising, ways. The present collection offers vivid evidence from among Father Busa's own publications of his ambition to build a worldwide network of scholarship in the interdisciplinary field he was helping to create: literary (or linguistic) data processing.

Steven E. Jones

Steven E. Jones is DeBartolo Chair in Liberal Arts and Professor of English and Digital Humanities at the University of South Florida. He is Project Director for "Reconstructing the First Humanities Computing Center", supported by a major Level II Digital Humanities Advancement Grant from the NEH (2017-2019). He founded and coordinates USF's DHLabs, a shared space for collaborative research in the College of Arts and Sciences. Before coming to USF in 2016 he was Distinguished Visiting Professor at CUNY Grad Center in New York (2014-2015) and taught for 28 years at Loyola University Chicago, where he co-founded and co-directed the Center for Textual Studies and Digital Humanities. He is author of numerous essays and books, including *Roberto Busa, S.J., and the Emergence of Humanities Computing* (Routledge, 2016) and *The Emergence of the Digital Humanities* (Routledge, 2014).

References

Busa, Roberto. Unpublished autobiographical manuscript. (Cited with the kind permission of Marco Passarotti, CIRCSE).

Jones, Steven E. 2016a. *Roberto Busa S. J., and the Emergence of Humanities Computing: The Priest and the Punched Cards*. Routledge.

Letter from Arthur K. Watson to Roberto Busa, April 7, 1960. Busa Archive ([14] CAAL ADDENDUM–[1] Primo raggruppamento (donazione sacerdote s.n.)–[4] CAAL Documenti).

Letter from Roberto Busa to Robert D. Eagleson, July 4, 1966. Busa Archive (Rel. Cult. 1944- Misc.).

Preface and Acknowledgements

Fr Roberto Busa S.J. (1913–2011) is often described as the founding father of humanities computing (now often called digital humanities)[1]: "Most fields cannot point to a single progenitor, much less a divine one, but humanities computing has Father Busa, who began working (with IBM) in the late 1940s on a concordance of the complete works of Thomas Aquinas" (Unsworth 2004). Yet, when perusing the secondary literature on Busa, it can seem that the total number of publications that closely analyse Busa's scholarship is inversely proportional to the total number of publications that broadly evoke his achievements and founding father status. That the secondary literature also contains a number of sweeping claims about Busa's work and context can hardly be unrelated to this. Fraser, for example, was apparently unaware of the centrality of concordances to the humanities, and of how they were obvious candidates for mechanization,[2] when he asked: "who would have been interested in concordances and indexes if Fr Busa had not made the connection between Aquinas' Latin style and the computer's innate ability to count?" (2000, 269) Yet, as Busa's 1951 publication shows, by the time he began his work concordances and indexes were long established and primary tools for teaching, learning and researching the humanities (see Chap. 2). To a large extent, Busa could not have chosen a more conventional form of scholarship to pursue,[3] and Fraser's implication that humanities computing would not have worked on concordances were it not for Busa is unconvincing.[4]

In the quote above, Fraser also implies, as have others, that Busa worked with computers from the outset. He did not. As the articles included in this volume attest, for much of the first decade of his research on the *Index Thomisticus*, Busa and his team used electromechanical accounting machines to encode and process the text of Aquinas and related authors.

---

[1] We tend to use digital humanities rather than humanities computing in this text because the former has gained particular traction since c. 2004 (see Kirschenbaum 2010; Rockwell and Sinclair 2016, 73–4). When we use humanities computing it is usually to refer to the pre-2004 period of the field now known as digital humanities. As made explicit in the title of this book, we view Busa's work as having given rise to one strand of humanities computing and acknowledge that other genealogies exist and are of crucial importance for understanding the emergence and development of the field (see, for example, Earhart et al. 2017)

[2] As Oakman observed: "Since concordance making involves several basic elements of data processing, it is not surprising that this literary application was the first one which received wide computer assistance" (Oakman 1973, 412)

[3] For an outline of the c. 700 year history of concordances see Raben (1969); for the early history of automated concordances see Burton (1981).

[4] Concordances were also of interest to fields like Machine Translation from the 1950s at least, see for example, Booth et al. (1958) and Vanhoutte (2013).

Moreover, as Jones has recently shown: "The application of this data-processing technology to linguistic research was really only proleptically and obliquely related to the humanities computing that would emerge (and be constructed) in the years that follow" (Jones 2016, 5).

The secondary literature on Busa also includes anachronistic claims about his work and intentions. For example: "[t]he first electronic text project in the humanities began in 1949 when Roberto Busa started work on his *Index Thomisticus*" (Hockey 2000, 5).[5] The work that Busa was doing in 1949 neither was, nor claimed to be, an electronic text project (see Chap. 2). In the 1950 announcement of his work in *Speculum,* Busa indicated that his aim was to create a file of word slips (such as were commonly used in dictionary making). His model was the "preliminary file used in preparation of *Thesaurus Linguae Latinae*" (Busa 1950, 425). He hoped mechanization could deliver the "greatest possible accuracy, with a maximum economy of human labor" (Busa 1950, 425). Far from electronic text, Busa initially contextualized and communicated his work with reference to analogue processes of dictionary and wordlist making.

Towards the end of the 1950s, Busa did discuss the manipulability of text that his work facilitated and these discussions include references to what might be described as antecedent or constitutive features of electronic text. For example, he wrote how "the new method, at half the price required for the preparation of the printing of a Concordance, gives not only the matrices for printing but also the entire catalog *in a flexible form always ready for new studies*" (see p. 48; emphasis ours). Yet, in that article, Busa envisages that the output of those new studies will be printed texts or the punched cards that lead to new printed texts. And so it had to be. Those technological developments like personal computers, networked computing and graphical user interfaces that would underpin electronic texts were still many years away. In the early 1960s, Busa does start to use terms like "magnetic book": "Books and manuscripts will remain, and currently the "magnetic book" takes its place by their side" (see p. 84). Yet, he does not there unpack this concept in sufficient detail to establish how his idea of a "magnetic book" relates to that of an "electronic text". Again in 1964, for example, he wrote that one of the problems that continued to occupy him was how to print a concordance that would occupy "500 volumes of 500 pages each. We are making an experiment for adopting a kind of microprint readable by means of a magnifying glass to be placed on a book and to be moved only downwards" (Busa 1964, 77). In summary, Busa was

---

[5] Hockey defines an "Electronic text in the humanities" as having the following characteristics: it is "an electronic representation of any textual material which is an object of study for literary, linguistic, historical or related purposes" (2000, 1). It follows from the discussion that for a text to become electronic it should be "modelled effectively on a computer" (2000, 2). Ideally, the same electronic text should meet diverse research requirements and should adequately represent the "complex features" of humanities texts (2000, 3). It can be "searched and otherwise manipulated by computer programmes in many different ways" (2000, 3).

not at work on electronic text in 1949 and the shape of the trajectory from his work to the electronic texts of later periods is incompletely understood.

In making the above points our aim is not to pedantically nit-pick. As Mahoney has written:

> When scientists study history, they often use their modern tools to determine what past work was "really about"; for example, the Babylonian mathematicians were "really" writing algorithms. But that is precisely what was not "really" happening. What was really happening was what was possible, indeed imaginable, in the intellectual environment of the time; what was really happening was what the linguistic and conceptual framework then would allow. The framework of Babylonian mathematics had no place for a metamathematical notion such as algorithm (Mahoney 1996, 831–2).

Following Mahoney, we believe that inaccuracies and anachronisms like those discussed above do matter. They point to an incomplete understanding of Busa's work and legacy. They also point to the necessity of studying Busa's contributions in their own terms and, as far as possible, in their actual historical context rather than that of twenty-first century humanities computing or digital humanities. Indeed, this observation was the jumping off point of this project. With this volume we hope to contribute to the project of building better understandings of what Busa thought he was "really" doing. Of course, one should not approach Busa's writings naïvely. They do not offer a neutral window on to his work; they must be read with the same caution and critical orientation as any other historical document. Yet, without better access to his published writings, and the possibility of bringing them into conversation with other sources that this will open, our efforts to better understand and contextualize Busa's work will not have a firm footing.

Despite the importance of Busa's work to understanding the emergence and development of fields like humanities computing and digital humanities, a large part of his *oeuvre* has remained inaccessible, or difficult to access, until this book. Many of his publications are either out of print or included in conference proceedings that had limited circulation and are now available in a few geographically dispersed libraries only. Also, Busa published in many languages, including German, French, Portuguese, Hebrew, Latin and Italian. Many humanities scholars will be able to read a few of these languages but not everyone can read them all. In this volume we make selected and translated writings of Busa available once more; many appear here in English for the first time.

A number of criteria informed our decisions about the texts that we have included. We aimed to include mostly out-of-print publications or publications that are otherwise difficult to access. We also aimed to include a representative selection of the topics that Busa addressed in his writings: technical, linguistic and philosophical. The process of translating the articles, and working them into the form they now have, was a long and unexpectedly difficult one. Busa's writing style is dense and metaphor-rich and this alone made his articles difficult to translate. Other problems were raised by the technical, synchronic and domain-specific terms that are used in his writings. We were not always certain about the most appropriate translation of those terms because they can refer to technologies, concepts and disciplines that are now obsolete. When we remained unsure of the most appropriate translation we supplied the term used in the original article in footnotes. Some writings also contain terms and features that are less acceptable to modern readers, for example, the ableist "*Hochgeschwindigkeittrottel*" (high speed cretin). The ostensible absence of women from the operations that Busa describes, even though we know this to not actually have been the case (see Nyhan and Terras 2017), is also problematic. After careful thought we decided to keep the translations as close to the originals as possible. Busa was a man of his time and place and it is not our task to hide this (or to presume that we are any less of ours). We do, however, provide a point of qualification in some of the "Editors notes" that stand at the head of each chapter where we thought it appropriate.

The process that led to the translations that are included here went as follows: scans were made of the original texts that are stored in the Busa Archive of the Library of Università Cattolica del Sacro Cuore, Milan. The scans were OCRed and checked. Next, the files were sent to the translators who had agreed to work on them. Once the translations had been returned to us we proceeded to work through each text at least two times, checking the translations and attending to questions about domain specific language, for example. At that point we decided to exclude some of the texts we had initially selected and we finalized our selection for this book. We regularly consulted our colleagues and incorporated many of their corrections and suggestions into the working translations (any errors that remain are ours, of course).

The vast majority of the articles included in this book were translated by Philip Barras, who worked with Busa for years and called him a friend. Even though Busa spoke and read a number of languages we suspect that he worked with many translators over the course of his career. Barras is one of the few translators with whom Busa openly acknowledged having worked.[6] So as to foreground the care and knowledge with which Barras translated Busa's work for this volume, and to record his recollections of having worked with Busa, we also carried out and include an oral history interview with Barras (see Chap. 20). We wish to thank Barras most sincerely for the trojan work that he did on these texts and for the care and conscientiousness he brought to his task.

Thank you also to Tessa Hauswedell (Chapter 5) and to Andreia Carvalho (Chapters 13 and 16) for the excellent translations they provided. We are also indebted to Geoffrey Rockwell for his exceptional contributions to Chapter 10 and for the help and guidance he gave us during this project. We have benefited immensely from his expertise and collegiality. Additional editorial assistance was provided by Marinella Testori, Jessica Salmon and Qin Lin, for which we are grateful.

[6] In the bibliography that Busa drew up he acknowledges two other translators: M. Nicolodi and E. Riccato (see Chap. 19).

We are also indebted to many other individuals and organizations for the diverse support they gave this volume. Without the philanthropy and kindness of Cristiana Costa this volume would not have been possible. Supplementary financial support was also secured from the Centre for Critical Heritage at the University of Gothenburg, Sweden and UCL, the Department of Information Studies UCL and the Faculty of Arts and Humanities, UCL.

Throughout this project, as indeed through many other projects, we have been shown immense kindness by Paolo Senna, Librarian at the Università Cattolica del Sacro Cuore. We thank him and hope we can benefit from his expertise and calm enthusiasm for many years more. Thank you also to Paolo Sirito, Director of the library of the Università Cattolica del Sacro Cuore and to Savina Raynaud, former Director of the CIRCSE Research Centre, Università Cattolica del Sacro Cuore. The assistance of Gian Luigi Brena S.J. and Roberto Gazzaniga S.J. from the Aloisianum, Gallarate and also of Danila Cairati (the final secretary to Busa) also deserves mention. We thank Willard McCarty, who first suggested that a book of translations of the work of Busa would be a boon for those who research the history of digital humanities.

The Society of Jesus is the copyright holder of the materials that are included in this volume. We secured permission to print translations of the articles contained in this volume from them; we are most thankful for their generosity and foresight. Thank you in particular to Maria Macchi of the Society of Jesus who expedited our requests so impressively. In addition to this we also contacted numerous editors and publishers of Busa's scholarship about this volume, where necessary also securing rights to reprint translations from them. We have made every effort to trace copyrights to their appropriate holders. If we have inadvertently failed to do so properly we apologize and request that they contact the publisher.

Most of all, we must thank Arianna Ciula, who made an immense contribution to practically every stage of this project. The field of digital humanities is made all the better by the kindness of colleagues like Arianna Ciula and those mentioned above—thank you.

Julianne Nyhan & Marco Passarotti
June 2019

References

Booth, A.D, L. Brandwood and J.P. Cleave. 1958. *Mechanical Resolution of Linguistic Problems*. London: Butterworths Scientific Publications.

Burton, D.M. 1981. Automated Concordances and Word Indexes: the fifties. *Computers and the Humanities* 15(1): 1–14.

Busa, R. 1950. Announcements. *Speculum* 25(3): 424–5.

Busa, R. 1965. An Inventory of Fifteen Million Words. In *Literary Data Processing Conference Proceedings September 9,10,11 1964*, ed. Jess B. Bessinger, Stephen M. Parrish, and Harry F. Arader, 64–78. Armonk: New York: IBM Corporation.

Earhart, A., Jones, S., McPherson T., Ray Murray, P. and Whitson, R. 2017. Alternate Histories of the Digital Humanities. Panel presented at Digital Humanities 2017, Montréal, Canada.

Fraser, M. 2000. From Concordances to Subject Portals: Supporting the Text-Centred Humanities Community. *Computers and the Humanities* 34: 265–278.

Hockey, S.M. 2000. Electronic Texts in the Humanities: Principles and Practice. Oxford: Oxford University Press.

Jones, S.E. 2016. Roberto Busa, S. J., and the Emergence of Humanities Computing: The Priest and the Punched Cards. New York; Oxon: Routledge.

Kirschenbaum, M.G. 2010. What is Digital Humanities and What's it Doing in English Departments? *ADE Bulletin* (150): 55–61.

Mahoney, M.S. 1996. What Makes History? In *History of programming languages II*, ed. Thomas J. Bergin and Rick G. Gibson, 831–2. NY: ACM Press.

Nyhan, J. and M. Terras 2017. Uncovering 'Hidden' Contributions to the History of Digital Humanities: the Index Thomisticus' Female Keypunch Operators. Paper presented at Digital Humanities 2017, Montréal, Canada.

Oakman, R.L. 1973. Concordances from Computers: a Review. In *Yearbook of the American Bibliographical and Textual Society*, ed. J. Katz, 3:411–25. Columbia: University of South Carolina Press.

Raben, J. 1969. The Death of the Handmade Concordance. *Scholarly Publishing* 1(1): 61–69.

Rockwell, G. and S. Sinclair. 2016. *Hermeneutica: Computer-Assisted Interpretation in the Humanities*. Cambridge, MA; London, England: The MIT Press.

Unsworth, J. (2004). Forms of Attention: Digital Humanities Beyond Representation. Paper delivered at The Face of Text: Computer-Assisted Text Analysis in the Humanities, the third conference of the Canadian Symposium on Text Analysis (CaSTA, McMaster University, November 19–21, 2004. http://people.lis.illinois.edu/~unsworth/FOA/ (accessed 17 March 2019).

Vanhoutte, E. 2013. The gates of hell: history and definition of Digital | Humanities | Computing. In *Defining Digital Humanities: A Reader*, ed. M.M. Terras, J. Nyhan, and E. Vanhoutte. Surrey: England; Burlington: USA: Ashgate Publishing Limited.

About the editors

Julianne Nyhan is associate Professor of Digital Information Studies at UCL (University College London), where she leads the digital humanities MA/MSc programme. She is also Deputy Director of the UCL Centre for Digital Humanities. Nyhan has published widely on the history of Digital Humanities, most recently (with Andrew Flinn) *Computation and the Humanities: towards an Oral History of Digital Humanities* (Springer 2016). She is a co-Investigator of a Leverhulme-funded collaboration with the British Museum on the manuscript catalogues of Sir Hans Sloane (https://tinyurl.com/y7zvrthm); a UK Principal Investigator of a digging into data challenge 'Oceanic Exchanges: tracing global information networks in historical newspapers' (http://oceanicexchanges.org/); and a co-Investigator of a Marie Curie action 'Critical Heritage Studies and the Future of Europe' (http://cheurope-project.eu/).

Marco Passarotti is associate Professor of Computational Linguistics at Università Cattolica del Sacro Cuore (Milan, Italy), where he is Director of the CIRCSE Research Centre. A former pupil of Fr Roberto Busa S.J., since 2006 he has headed the *Index Thomisticus* Treebank project, which continues the legacy of Busa's work on the *opera omnia* of Thomas Aquinas (https://itreebank.marginalia.it/). He is the Principal Investigator of the *LiLa* project (https://lila-erc.eu/), an ERC-Consolidator Grant (2018–2023) which aims to build a Linked Data Knowledge Base of linguistic resources and natural language processing tools for Latin. He co-chairs the series of workshops on 'Corpus-based Research in the Humanities' (CRH).

Chapter 1 Introduction, or Why Busa Still Matters

Marco Passarotti and Julianne Nyhan

Introduction

Father Roberto Busa S.J. did not choose to become a scholar. As he recalled it, the decision was made for him. In 1933, at the age of twenty, driven by his vocation to become a missionary, he joined the Society of Jesus. Shortly after being ordained in 1940, Busa came before his superior who had the task of assigning him to an area of expertise within the Society. Busa often recalled that moment in the form of a dialogue:

> [Superior]: "Would you like to become a professor?"
> [Busa]: "In no way!" My wish was to be a missionary to take care of the poor
> [Superior]: "Good! You'll do it, all the same" (Busa 1980, 83).

And so he was sent to the Pontifical Gregorian University in Rome where, in 1946, he was awarded a degree in Philosophy for a thesis entitled *La terminologia tomistica dell'interiorità*, which would later be published as a monograph (Busa 1949).

Busa may initially have been a reluctant scholar; yet between 1949 and 2009 he published in the region of 350 scholarly contributions.[1] His publications ranged across many subjects but often addressed topics in the domains of philosophy, theology, computational linguistics and humanities computing. The texts included in this volume alone discuss the electromechanical and computational techniques that he and Paul Tasman developed for the *Index Thomisticus*; articles about the application of computing to language and philology; philosophical writings on humans and computers; and concordances and lexicostatistical analyses of texts in Latin (and other languages). In these publications we find references to topics and technologies that now sound quite dated or have become obsolete, for example, cybernetics, punched card machines, electronic calculators and CD-ROMs etc. In some ways, to read Busa's articles is to understand Gange's observation (that was made about the history of Egyptology but has wider applicability) that "the gulf between the scholar in the present and the Egyptologist of even fifty years ago is far wider than is commonly assumed" (Gange 2014, 64).

[1] It is difficult to give a precise number because Busa's texts were often translated and republished (see Chap. 19).

What, then, is the relevance of Busa's twentieth-century publications to the twenty-first century fields of digital humanities, computational linguistics and beyond? Why and how does his work still matter? We argue that Busa's methodological approach remains valid, despite the unceasing ebbs and flows of tools, technologies, formats and disciplinary boundaries. As we shall explore, Busa's approach was founded on the belief that humanities research should not be impressionistic, or based on selected examples, but that any interpretation should be based on all the data available to support it, thus allowing for replication of results. Busa's methodological approach has not become old, but still remains (and must remain) a keystone of many kinds of computational work in the humanities.

So too, we argue that Busa's publications are crucial sources for writing the histories of digital humanities, and thus for understanding the present shape of the field and imagining its futures. The project of writing the histories of digital humanities is a necessary and urgent one. As McCarty has written:

> Digital humanities needs [to] use its 64 years of fumbling to gain leverage for a great inductive leap to a vantage point from which its disciplinary shape and trajectory … can be clearly seen. The key to its future—and in some measure the future of all the related humanities—is its history. This history we must remember (McCarty 2014, 295).

To build a case for the continuing importance of Busa's publications we proceed by undertaking a review of some of distinctive themes found in Busa's individual articles and in the accretions of discussions that are sustained across them. We also draw these themes into conversation with some current thematic, theoretical and methodological concerns of digital humanities and computational linguistics and find much that still resonates. Marco Passarotti knew Busa personally and worked closely with him for many years. We have accordingly integrated into this text some details of conversations that Passarotti recalls where we felt that they could assist in the interpretation of the texts discussed below.

We proceed by discussing the following themes: the spiritual in Busa's writings; the computer and the humanities; what distinguishes humanities computing?; and speed versus research trajectories. Before concluding, we also discuss some of the new questions about Busa's work that are suggested by the articles that have been translated for, and are assembled in this volume.

The spiritual in Busa's writings

Busa taught for many years at the Pontifical Gregorian University in Rome and at the Università Cattolica, Milan, where he also set up the research group GIRCSE (Gruppo Interdisciplinare di Ricerche per la Computerizzazione dei Segni dell'Espressione, now called CIRCSE).[2] Yet he remained somewhat of an outsider of the

[2] See https://centridiricerca.unicatt.it/circse_index.html (accessed 18/06/2019)

Academy for much of his scholarly career. Busa did not hold a permanent academic post in a University[3] and he was, first and foremost, a priest. This is starkly evinced by the references to spirituality and religion that frequently appear in his scholarly *oeuvre*. These references can strike the reader as odd and it can be tempting to dismiss them as curious intrusions from Busa's spiritual life into his scholarly work. We argue, however, that they are important keys that can help to unlock deeper understandings of Busa's work and his particular *weltanschauung*. They are also relevant to ongoing discussions of how Busa's Jesuit context framed his work (see Jones 2016, 15–6), and thus, of how institutions outside of the University context may have shaped the earliest forays into digital humanities of which we are currently aware (see Nyhan and Flinn 2016).

Busa's writings and projects show that his life and work were strictly bound. He was always a scholar *and* a priest: those two roles could not be divided. As he wrote: "A Jesuit may be assigned to scholarship to become a specialist in any particular field, so that in a secularized world he may document scientifically that prayer is the logical continuation of the principles behind any branch of learning" (Busa 1998, 4). Busa recorded three kinds of information in his diary every day: his location, the names of those he had met during the day and the names of those for whom he had recited the Holy Mass.[4] Working and praying were his everyday life. He used to say that he had become a (computational) linguist not despite being a priest, but because of being priest and it was through the lens of a priest that he often viewed computing. In 1966, for example, he wrote how:

> The "exits" towards the recognition of the presence of God are remarkable and impressive: and precisely because information theory, science of government, and cybernetics are essentially nothing if not the analysis of the phenomenon of active organization, examined in its downward progress when it should be the other way around. That is, from the result towards the first dynamic principal, how is it not possible to understand immediately that all the complex periphery nonetheless always has a centre, and one only, which is its motive force, and to be its motive force can it not also be its inventor? … (see p. 101–2).

This strict connection between life and work, where one motivates the other in an iterative cycle, distinctively framed Busa's interpretation of the significance of the application of computing to language (and the humanities). Thus, his discussions of the significance of problems that were encountered in his work often gave way to discussions about God. He saw the difficulties that are encountered while trying to formalize even simple linguistic facts for processing as more than technical or linguistic problems. Busa argued that there was something more going on and that the steady confrontation with empirical data pointed to deeper mysteries:

[3] The Aloisianum, where Busa was professor and librarian in the Faculty of Philosophy, was not a university but a Jesuit institute.

[4] This information is drawn from the personal recollections of Marco Passarotti. Busa's diary is unfortunately not in his archive in the Università Cattolica del Sacro Cuore, Milan, and is believed to have been thrown away when he died.

> The automation of written language awaits some technical development, but it also expects much more from the spiritual industriousness of mankind. The machine warns us that we are not humanistic enough and, although we speak, we are not able to explain how we speak. It is the despised machine that repeats to us the invitation "know thyself still more profoundly, scientifically and humanistically: study your speech". The automation of written language thus promises an increase in spiritual education (see p. 68).

The line of reasoning discussed above, where Busa draws attention to what the computer *cannot* do, reflects on how this relates to the limits of human knowledge and sets out the insights that can flow from this observation, is one that he often followed.[5] In this book we see Busa emphasizing that the computer does not have innate intelligence (see Chap. 9); that it cannot "know" but only store information (see Chap. 12); that it cannot be a programmer (see Chap. 15); and that it cannot be produced by nature (see Chap. 17). Perhaps most famously, Busa contributed a guest editorial to the *Bulletin* of the Association for Literary and Linguistic Computing entitled "Why can a Computer do so Little?" (Busa 1976). We also see him building on these observations as he poses fundamental questions about what it means to know (see Chap. 12), to think, act and communicate (see Chap. 15), to use, understand and communicate (see Chap. 16) and to be human (see Chap. 17). Thus, what might be thought of as a negative approach (or one that pays particular attention to points of failure, difficulty and disruption in the encounter between human knowledge and computing) brims with potential because of the deeper questions it can raise, like "what is in our mouth at every moment, the mysterious world of our words" (Busa 1976, 3).

Though not usually with recourse to the explicitly faith-based dimensions that often framed his analyses, Busa's emphasis on the heuristic potential of the failure and difficulty that can occur at the intersection of computing and human knowledge arguably has proven influential among digital humanities scholars. Echoes of his approach can, for example, be detected in McCarty's seminal contributions to the theory of modelling in digital humanities (see McCarty 2005). For the purposes of this chapter we will describe a digital humanities model as an abstracted digital representation of an "object" of study (see e.g. Ciula, A, Eide, Ø, and Sahle P. 2019; Flanders and Jannidis 2018). Usually the features of an object that a researcher wishes to study, for example, rhyme or prosody, are emphasized in a model and made manipulatable by and through it. To realize this the researcher must first identify and describe those features with the complete clarity, consistency and explicitness that computing requires, something that can be difficult and sometimes impossible to do for works of imagination and learning. Paradoxically, then, McCarty has argued that the greatest successes of modelling are to be found in its failures, or its "*via negativa*". This gives us, he argues, "a tool for isolating that which will not compute and thus forces the epistemological question of how it is that we know what we really know in the humanities" (McCarty 2008, 256). In other examples of digital humanities scholarship that explore the role of

[5] McCarty has argued that Busa implicitly followed "Turing's use of the machine to illumine what it could not do" (2013, 4).

tension, defamiliarization and deformation in furthering critical interpretation and engagement we can also detect a reverberation of Busa (for example, McGann 2004; Ramsay 2011). Flanders, for example, has written on the "productive unease" evoked by digital scholarship: "This unease registers for the humanities scholar as a sense of friction between familiar mental habits and the affordances of the tool, but it is ideally a provocative friction, an irritation that prompts further thought and engagement" (Flanders 2009).

The computer and the humanities

Busa's writings also include discussions about the role of computing in the humanities and whether the computer could make the humanities obsolete. In exploring these questions Busa began to articulate what he believed to be distinctive about humanities computing research and he identified some of the wider projects that this research could inform. These topics are of enduring concern to present-day digital humanities. Busa's writings are thus important sources for understanding the longer history, and development, of these discussions and debates.

In 1962, Busa used an arresting metaphor for the reaction of some in the humanities to the advances that had recently been made in automation: "At this point a nightmare intervened, technology triumphant with its latest creation: automation. People shuddered, considering it a crude, hard bulldozer that goes roaring ahead, crushing and shredding flowers, amongst which, a delicate and gentle victim, is humanism" (see p. 79). Just three years earlier, Snow had published his now famous treatment of the differences and mistrust he saw between the two sides of the scholarly world: "two groups, comparable in intelligence, identical in race, not grossly different in social origin, earning about the same incomes, who ha[ve] almost ceased to communicate at all" (Snow 1959, 2). Instead of the mutual disregard mentioned by Snow, Busa speaks of the fearful, even aggressive, reaction of humanists to automation. He portrays them as a group who believe themselves to be victims of a methodological revolution founded on a reductive instrumentalism. He also implies that humanists attacked automation in this way so as to deflect from their embarrassment at the new questions it raised that they could not answer:

> Tomorrow is already upon us. The future has already begun [...] the men involved in automation began to [...] ask philologists and grammarians, who were busy in the fields selecting the choicest flowers, questions such as these: Please, how many verbs are there in Russian that are active and transitive, and how many that are active and intransitive? How many are there in English? [...] Please, would you arrange all the words in the dictionary according to the various morphological and grammatical categories? Would you please tell me which words may be omitted, and when, so as to shorten a text without any detriment to its meaning? (see p. 79).

What Busa calls "tomorrow" is the computational processing of textual data, which demands a comprehensiveness of linguistic knowledge that humanists did not, and perhaps could not have had in 1962. The questions that Busa puts to humanists from the "men of automation" concern research topics that, in some cases, could have been explored at scale only in the decades after his paper, as digital corpora of the relevant languages became available. The first question is about the transitive/intransitive use of verbs. To ask such questions today, we use syntactically annotated corpora (or treebanks), which were not available in 1962. As for the second question, on "morphological and grammatical categories" of words, at the time of writing we answer this with natural language processing tools like Part of Speech (PoS) taggers or morphological analysers. The third question has been responded to in recent years through lines which have seen large growth in response to the needs raised by the internet, like text summarization and key-word extraction.

Busa's use of the bulldozer analogy and his emphasis on humanists' inability to answer the questions raised in the course of formalizing language could be taken to imply that he viewed the humanities as moribund: "a machine made us realize that no humanist has such command of his own language as to be able to answer such questions. A machine [...] has revealed that there is still too little humanism of the serious and systematic type" (see p. 79). Yet, as he argued elsewhere, automation not only foregrounds these problems, it also offers a means of pursuing them: "Not only do computers invite us to wider, deeper, and more systematic research, they also make it possible" (see p. 89). Busa argued that the limitations brought to light by a machine could be used by the humanities to make a momentous step forward. The required methodological turn could raise a new kind of research in the humanities, founded on an exhaustive and systematic approach to linguistic data:

> Automation of the treatment of information requires the automation of the compilation of indices, concordances, and of all the possible types of statistics of linguistic facts. [...] you will realize that a new lexicology and new linguistics into techniques for the treatment of information are developing amongst the researchers. This lexicology and linguistics is more systematic, more exhaustive, more widely useful, and, I am emboldened to say, more humanistic than the traditional ones in use up to now (see p. 81–2).

So too, it would bind the humanities to those fields that addressed questions of natural language processing, including those which worked on the high-priority economic, defence and security issues of his day. In the following, for example, it is worth noticing that Busa mentions the "activities of production, exchange and defence" as the ones motivating automation in the area of information retrieval. Those were the years of the so-called "Italian economic miracle" and the Cold War:

> Economic facts today demand a qualitative increase of grammatical and lexical sciences as one of the necessary conditions of their vital development. ... The activities of

production, trade, and defence demand the automation of "information retrieval", which I would translate as an opportune system for the tracing of useful knowledge (see p. 79).

In this way Busa can be seen to make the case for the ongoing, and in fact, increased relevance of the humanities in the age of automation. It is notable, however, that he makes this case without addressing the ethical questions that are raised by the proposed association of the humanities with the military-industrial complex. The ongoing relevance of the humanities is a topic to which he would again return many times, for example, in his Busa award acceptance lecture: "I repeat: computerized speleology, to retrieve deep roots of human language, is fundamental in all disciplines. At this level, humanities are the prime source and principle for all sciences and technologies" (Busa 1999, 7).

What distinguishes humanities computing?

In the 'bulldozer' article above we saw Busa claim that humanists were busy "selecting the choicest flowers", or picking up selected samples of evidence only. In this highly critical expression there is much of Busa's thought, whose core position was that research in the humanities should reach inductive conclusions only from exhaustive empirical data. He saw this as the fundamental contribution of computationally-mediated research and a desideratum of pursuing it: "the inductive interpretation of the phenomenon of language [...] promises [...] to restart the cycle of linguistic and grammatical awareness with greater depth, methodicalness and documentation" (see p. 84). This is particularly evident in the approach that Busa took to the processing of function words. As he pointed out: "an important scientific role is played by [the] processing of function and high-frequency words (pronouns, *et, non, sum*, etc.); this was almost never done previously because it is infeasible manually, but it is practical using a computer" (Busa 1980, 87). Thus, the *Index Thomisticus* project recorded and analysed even "*et*" (and).

Busa was insistent that neither selected samples nor human intuition alone could validate a linguistic hypothesis. He argued that the use of computers to process large amounts of linguistic data would in turn raise the quality and reproducibility of experiments, thus enhancing the scientific degree of the humanities. Discussing queries that were run on non-lemmatized wordforms, for example, he wrote: "I cannot consider "scientific" the final documentation produced by such research methods. This will always provide only rough and impressionistic data: aren't there already enough in academic production and especially in the humanities?" (Busa 2000, 167; translation Passarotti).[6] In that same text he emphasized the close link he held to obtain between "scientific" and "empirical", i.e. "induc-

[6] In the original: "...non mi sento di ritenere scientifico il documento conclusivo di tali modi d'indagine [...]. Esso fornirà sempre dati soltanto approssimativi e di opinione: non ve ne sono già abbastanza nella produzione accademica, specialmente nelle scienze umane?".

tive" and not only "deductive". He wrote: "I claim that empirical can have two meanings: one of "not scientific" and the other of "scientific", but achieved (also) after experimentation and observation and not only with deductive reasoning" (Busa 2000, 116; translation Passarotti).[7] Elsewhere he claimed that "Far from diminishing humanism in any way, computers actually promote our humanism to the perfection of a scientific method" (see p. 89).

The idea that the humanities would or should be made more scientific is one that many scholars would rightly push back against. From our reading of Busa's texts we have concluded that in using the term "scientific" (*scientifico*) he was using it in the broad sense of *wissenschaft,* or the systematic pursuit of knowledge that is not necessarily tied to any particular discipline. With this term it seems that he also sought to evoke the idea of replicability in the humanities. When describing his own work, Busa often sought to specify the linguistic information that could help the reader to repeat the work that he had done. For example, he described in detail the steps that were taken to organize the lemmas of the *Index Thomisticus* into "types of semanticity" (Busa 1994).

As was his habit, Busa often communicated his ideas to colleagues with a metaphor. He would remark that most research in the humanities is like a mile of algorithms on a mere inch of foundation. He contrasted this with the methodology he employed throughout his research life. On a foundation a mile long, he sought to raise the research by an inch along the whole length of the mile. He then sought to raise the level by another inch along the whole mile, and so on. All the evidence provided by each level of analysis was taken into consideration before moving on to the next level, which was slightly more advanced than the last (see also p. 142; Busa 1990). According to Busa, only in this way was it possible to provide a solid basis for research conclusions.

Among the flurry of activities and research questions raised by the automatic processing of linguistic data, Busa emphasized the fundamental aspects of his research:

> My contribution [...] deals with [...] the development of operational methods that permit research into the first numerical proportions intrinsic to language. [...] I am engaged in working out techniques that allow one, rapidly and on a large scale, to isolate, calculate, and codify the presence and proportions of frequency of words (distinguishing and separating inflections, homographs, compound words ...), morphemes (roots, prefixes, suffixes ...), syllables, letters and phonemes, accents, distribution of the parts of speech, length of sentences and phrases, etc. (see p. 66).

Not by chance, in those years the US government and military largely funded fundamental research in machine translation, which was much reduced after the ALPAC report (ALPAC 1966). This report found that before focusing on the problems of machine translation, fundamental linguistic research on the basic but

[7] In the original: "[...] opino che "empirico" possa aver due valori: uno di 'non scientifico', l'altro di 'scientifico', ma acquisito (anche) con sperimentazione od osservazione e non con soli ragionamenti deduttivi".

essential levels of linguistic analysis, like tokenization, lemmatization and PoS tagging was necessary.

Not speed but a research trajectory

Present-day researchers do not usually hold that achieving gains like efficiency or cost-effectiveness should be goals of computing in humanities research (see, for example, Prescott 2012). In the earliest publications included in this volume we nevertheless see Busa emphasizing the gains in speed, efficiency and value for money that mechanization could offer. For example: "even with standard machines the punched card system permits more extensive, more certain, more advanced and more economical studies than would have been possible except with the patient work of many men" (see p. 56). He also used analogies for the methodology he and Tasman developed that evoke concern for efficiency and cost-effectiveness, for example, "A slightly more distant programme is to carry out the work on something that contains four or five million words. Such an experiment corresponds to the well-known transition from laboratory tests to mass production" (see p. 66).

Yet, for the most part—perhaps because it was becoming clear to him that the completion times he had initially foreseen for the *Index Thomisticus* would be well surpassed (see p. 31)—Busa started to move away from such claims by the end of the 1950s. From then he tended to focus on the slow and painstaking nature of the work and on the longer-term gains that it offered in the quality and systematicity of results. He also turned his attention to discussing the contours of the research trajectory that was needed to vouchsafe these gains over the longer term. In 1958, he discussed the main problems met by machine translation, which was then the main area of research in the field of automatic linguistic processing:

> The main difficulties come respectively from the limitations of a machine's memory and the size and complexity of linguistic facts, together with the still rudimentary knowledge that we have of them: it is enough to think of the problems raised by homography, by the different inflections a word can have, by syntax, and by idioms and metaphorical expressions (see p. 64).

To overcome this, and to make it possible to write computer programs able to process linguistic data automatically, Busa identified a number of linguistic issues that needed to be studied more. He identified issues that would become core research areas of computational linguistics in the decades to come: morphological analysis, PoS tagging ("homography" and "the different inflections a word"), syntactic parsing and metaphor processing.

In 1983 he looked back at the research that had been done on these issues in the intervening years and delivered a critical assessment of the progress that had been made. In the following quote his reference to what was done "more than thirty years ago" is presumably a reference his work on the *Varia* (see Chap. 2):

> Linguistic informatics [...] is still re-cycling without superseding the qualitative leaps forward made more than thirty years ago. [...] firstly, ... there has been no general progression beyond the morpho-lexical categorization of the individual words, i.e. taken one by one. Secondly, ... the fierce attack on automatic lemmatization, syntactic-semantic analysis and syntactical-logical analysis has hardly begun (see p. 137).

Busa regretted that the elaboration of concordances had become an end in itself. Concordances were not being treated as corpora that could be exploited to support the processing of further levels of linguistic analysis (see p. 135). Consistent with his fully inductive method, Busa urged the community to move one step further, i.e. from morphological analysis of out-of-context words (types) to their contextual analysis (tokens), which would lead to developments like PoS tagging, syntactic parsing and semantic processing. In particular, Busa proposed a number of research topics that he considered to be essential for the development of the discipline and that would become pivotal in computational linguistics only years (or decades) later. Here are some of them:

> have you ever come across computerized tables and concordances in which the programme automatically re-unites names with their respective surnames? Or extracts the compound forms of verbs? Or connects pronouns with the nouns that they represent? Or marks the ellipses? (see p. 137–8).

Today, all this is performed by different natural language processing techniques, like named-entity recognition, chunking, syntactic parsing, anaphora and ellipsis resolution. For many languages, these tasks still provide low accuracy rates and their effectiveness is limited to specific domains (like medical terminology). But they are generally considered essential components in automatic textual analysis.

A few lines later, Busa sketches another important need: "to identify all the words that can be arranged (not all can) in hierarchical pyramids, from the more universal to the more specific" (see p. 140). In this passage, Busa seems to anticipate the research that, just in those years, George A. Miller was starting at Princeton University, i.e. the lexical resource named WordNet (see Fellbaum 2005). Today WordNet is available for a number of languages and is widely used by scholars as well as by industrial applications. And finally, here it is, PoS-tagging, metaphor processing and ellipsis resolution: "I still have not seen treatments that are inductively documented of the phenomenon of homography, nor of that of metaphor, nor of the vicariance of pronouns, [...] nor of implied words, those that are expressed by not expressing them" (see p. 140).

Busa never stops to confirm what he had already claimed in 1962, that the inductive analysis of linguistic phenomena would lead to a new awareness of them:

> I think that there is not one paragraph in our grammar books that, with one or more inductive researches on the computer, could not be finally documented and re-systematized in such a way as to be capable of being inserted into computer lists without causing too many unpleasant surprises (see p. 141).

He presents the rewriting of the basic sections of traditional grammars as the fundamental ground from which the discipline could take flight and glide to more complex levels of linguistic analysis, among which machine translation represented the highest peak:

> Automatic indexing and automatic abstracting are still dreams, and what little automatic translation is being carried out flows in the channel of contemporary technical-scientific writings. Yet, nonetheless, it is clear that linguistics would explode into an enormous industry of information if only these difficulties could be mastered. (see p. 138).

Busa clearly understood the importance of natural language processing, as a necessary step towards linguistics as "a huge industry of information". Today, for better or worse, the big names of communication (like Apple, Facebook and Google) are among the most important brands in the world.

When asked how he saw the future of computational linguistics, Busa used to reply that the discipline would experience a big boom thanks to increasingly powerful computers, the widespread diffusion of digital technology and the ease of transfer of information across the Internet. He foresaw that the wide availability of natural language processing tools, annotated corpora, lexicons and ontologies would run the risk of being incorrectly exploited. Yet he believed that the greatest danger lay in considering computational linguistics not as a discipline aimed at doing things better, but rather as a tool to do things increasingly faster.[8] He feared that the computational linguists of the third millennium would become picky about dealing with data (which should be their bread and butter) and lose the humility to check each analysis, preferring to process huge masses of texts quickly and approximatively, without even reading a line.

Busa was able to aim at huge long-term goals and, at the same time, to understand and rigorously implement the single (sometimes tedious) steps that were necessary for the *Index Thomisticus*. As we have seen, Busa had an intimate relationship with linguistic data. He managed data with absolute rigour and the motto "*aut omnia aut nihil*" characterized his entire scientific production. His rigorous, empirical and systematic approach to linguistic data should remain at the core of computational linguistic research and digital humanities, instead of allowing those fields to merely become hunting grounds for the best-performing natural language processing tool, or the largest annotated corpus.

Today, the work of Busa continues with the project of the *Index Thomisticus* Treebank (Passarotti, 2014), at the CIRCSE research centre of Università Cattolica del Sacro Cuore, Milan. This project began in 2006 and aims to produce the syntactic annotation of the entire *Index Thomisticus*. It has inserted the *Index* into the cutting edge of annotated corpora and linguistic resources for modern languages, making Latin, the mother-tongue of computational linguistics, a language which is

---

[8] Passarotti recalls many discussions about this with Busa but as far as we are aware he did not publish much on this topic. With reference to the so-called "Lingue Disciplinate" (for machine translation purposes), he wrote "Le procedure iniziali di tale processo impongono operazioni inizialmente anche umane, ripetitive per tempi lunghi. Esse non sono adatte a chi avesse scadenze a breve termine" (The initial procedures require time consuming operations that are, at the beginning, also manual. These do not fit with those who have short time deadlines) (Busa 2003, 62; translation Passarotti).

today no longer less-resourced. To this end, the *Index Thomisticus* Treebank enjoys close collaboration with other treebanks of ancient languages (above all, the Ancient Greek and Latin Dependency Treebanks (http://perseusdl.github.io/treebank_data/) and the PROIEL corpus (Haug & Jøhndal 2008)). It has recently been integrated into the CLARIN infrastructure of language resources (www.clarin.eu) and into Universal Dependencies, the large repository of treebanks (http://universaldependencies.org/).

New questions

Before concluding, we will turn to consider one of the many new research questions that are raised by the articles that are gathered here. When Busa began work on the *Index Thomisticus* in the early 1950s, concordances and indexes were authoritative and common information retrieval tools with a pedigree that stretched back to the thirteenth century (Wisbey 1962, 161; see also Blair 2010, 141–4). Technologies like the printing press had played an important role in the standardisation[9] of the index form (see Eisenstein 1979, 91; Pettegree 2011, 294). Yet the initial conceptualization and elaboration of indexes, concordances and other research tools had occurred in medieval manuscript culture (see Rouse and Rouse 1993, 255). In contrast with major works of twelfth century, which sought to "assimilate and organize inherited written authority in systematic form", the genre of reference tools (including concordances) that emerged in the thirteenth century allowed a text to be "used, rather than read" so that it was possible for the first time to "search written authority afresh, to get at, to locate and to retrieve information" (Rouse and Rouse 1993, 221).

Crucial to this was the development and use of reference systems, which allowed individuals to navigate from an excerpt in a concordance back to the text that it occurred in, so as to see that particular word or phrase in its full context. Manuscript dependent ways of doing this existed and involved, for example, "dividing up the physical manuscript itself according to numbered folios or numbered two-page openings, so that one may refer to page, column, even line" (Rouse and Rouse 1993, 243). For concordances to the Bible, it was necessary for the reference system to be "layout-independent, since each manuscript would vary in the amount of text included on each page" (Blair 2010, 38). The reference system that was developed for this used chapter numbering, and in due course verse numbering and systems of locating passages that relied on a reader mentally dividing a text into 7 sections, A-G (Blair 2010, 39; see also Rouse and Rouse 1993, 243–5). The literature on the reference systems that were developed for concordances and

[9] Eisenstein has argued that the reasons for this include the increasing use of full alphabetization to order written materials, "typographical standardization" and the "competitive commercial character of the book trade" (1979, 91).

indexes also mention some that were used to facilitate the indexing of a text (and the creation of semantic cross references between texts). An example of this is found in the early-thirteenth-century work of a team that was based in Oxford and led by Robert Grosseteste. They worked on patristic texts and devised

> a complicated series of symbols (Greek letters, mathematical and conventional signs, and so on) which a scholar could jot down in the margin of a work in the appropriate place while he was reading—with the ultimate goal of incorporating all the references into one integrated central index (Rouse and Rouse 1993, 232).

The system proved influential. The symbols continued to be used throughout the second half of the thirteenth-century and appear in at least 17 extant manuscripts (ibid).

Just as those teams of clerics, scholars and amanuenses who worked on the earliest concordances and indexes had to devise and execute appropriate reference systems so too did Busa. The discussions that are included in Busa's articles of the reference systems that he used are not especially detailed. Yet they do suggest that he used reference systems in line with those of the longer analogue concordance tradition. Broadly evoking the function of the symbols of Grosseteste mentioned above, during the phase of pre-editing the *Index Thomisticus*, Busa and his colleagues devised a system of symbols and markup that signalled the function of a given portion of text. For example, they used colour to distinguish "words quoted by the author from other writers, from the author's own words, etc" (see p. 43). This system facilitated the elaboration of the *Index Thomisticus* because it guided the keypunch operators who encoded the millions of words of the *Index Thomisticus* on punched cards (see Chap. 11). Symbols were also used on punched cards to record metadata about a word or phrase. For example, # was used to indicate "These are the words of another author whom St. Thomas quotes here literally" (Tasman 1957, 254).

Next to this, Busa also used various references and unique identification numbers to wrangle the millions of words that were being processed by the *Index Thomisticus* project. In 1958, he summarized this as follows:

> every word is coded as to its location with the reference and with the number of its position in the text; it is coded as a morphologic unit with the progressive number that it acquires in the first alphabetic sequence; it is coded as a semantic unit, with the progressive number that it has in the last alphabetical order (see p. 46).

The most important of the reference systems referred to in the quote above was arguably the line number reference because this allowed the reader to locate a particular word in the text it had been excerpted from (and, as in the example above of references to the Bible, the line number was presumably not dependent on the page numbers of different editions of the same text). The purpose of specifying each word's position in the text was that it would allow the complete text to be reassembled from the punched cards, where it was represented in atomized form (see p. 45–6). Words that were graphically identical were also assigned a sequence number and this facilitated the lexicostatistical calculations that Busa would later run on the text. A further reference number was also apparently also assigned to lemmatized forms of inflected words (see p. 45).

Though Busa published widely during his lifetime, and often discussed the technical methods that underpinned the *Index Thomisticus*, he usually did so at quite a high level of abstraction. It is accordingly difficult to reconstruct from Busa's publications a detailed account of the technical methods and processes that underpinned the *Index Thomisticus* (see Rockwell 2016). As a result, perhaps, there has been relatively little sustained and detailed study of how Busa's technical methods and processes unfolded during the 30 years that he worked on the *Index Thomisticus* and of how his method factor in the longer concordance tradition. Busa's publications show that he had a good knowledge of the history of concordances (see Chap. 2). To what extent, then, did his work incorporate the reference systems that had been devised for hard copy concordances? Was Busa's assignation of progressive numbers to morphological and semantic units innovative? Also, to what extent might Busa have drawn from, or contributed to, the reference systems that were devised or used by other communities, like the machine learning and library science communities, as they used electromechanical accounting machines and early computers to manage and retrieve textual information? (see, for example, Garfield 1955)

Perhaps by reading the articles in this book next to the documents contained in Busa's archive and following Sinclair and Rockwell (Sinclair 2016) in using approaches like humanistic fabrication to replicate the technology and techniques of the *Index Thomisticus,* it may be possible to understand more about the reference systems (and other methods and processes) that Busa used. A flow chart for Busa's "Mechanized Linguistic Analysis," that was produced at IBM, New York, in 1952, is one example of archival documentation that will help to shed some more light on the technical methods that Busa used in the earliest stages of the *Index Thomisticus* (Jones 2016b).[10] In the most immediate sense, this will lead to better and more detailed understandings of how the *Index Thomisticus* was executed. By contextualizing the outcomes of such a study with regard to the history of concordance making, including the history of reference systems, we will be better able to assess Busa's work and legacy within longer histories of the development of text-technology, philosophies of information and the histories of the humanities, including that of digital humanities.

Conclusion

In this chapter, we have explored the question: why does Busa's work still matter? In seeking to answer this, we have explored Busa's methodological contributions,

[10] The original document has not been catalogued at the time of writing; it is stored in the Busa Archive.

especially his position that research in the humanities should reach inductive conclusions from exhaustive empirical data. We believe that this insight continues to hold true for many kinds of research that are undertaken by the fields of digital humanities and computational linguistics. We have also shown that Busa's writings contain early articulations of questions that are still debated in the digital humanities. These debates address topics like: What is the role of the humanities in the digital age? What is distinctive about the digital humanities? What is the purpose of computing in the humanities? Busa's articles are thus important sources for developing better understandings of the trajectory of debates that have been central to digital humanities on many occasions over the past 70 years. We have also drawn attention to some of the questions that are raised by Busa's publications that we cannot currently answer. Our current understanding of how Busa's work built on, and whether it advanced, established methodologies and text-technologies (e.g. reference systems) of the analogue concordance tradition is not rich. Answers to questions like these are important because they may open better understandings of the place of digital humanities in the longer history of the humanities.

In the text above we have framed our discussion of the continuing relevance of Busa's work primarily in terms of the interests and perspectives of present-day digital humanities and computational linguistics. To close, we would like to return to those elements of Busa's thought that transcend particular historical moments and that we hold to be meaningful in and of themselves. In Busa's thought, lemmatizing or morphologically tagging a text requires and enables a ceaseless cycle of "know yourself". Today, these tasks are performed automatically for many languages. Yet his call to "know yourself" was the understanding of digital humanities that was most intimate to Busa. He saw in this challenge an endless source of new questions and a call to find new answers about what it means to be human. It is a challenge that is of timeless importance to the (digital) humanities.

References

ALPAC 1966. *Languages and Machines: Computers in Translation and Linguistics. A Report by the Automatic Language Processing Advisory Committee, Division of Behavioral Sciences, National Academy of Sciences, National Research Council*, Washington D.C., National Academy of Sciences, National Research Council.

Blair, A.M. 2010. *Too Much to Know: Managing Scholarly Information before the Modern Age*. New Haven; London: Yale University Press.

Busa, R. 1949. *La Terminologia Tomistica dell'Interiorità. Saggi di metodo per una interpretazione della metafisica della presenza*. Milano: Fratelli Bocca.

Busa, R. 1976. Guest Editorial: Why can a Computer do so little? *ALLC Bulletin*, 4(1):1–3.

Busa, R. 1980. The Annals of Humanities Computing: The Index Thomisticus. *Computers and the Humanities* 14(2): 83–90.

Busa, R. 1990. Informatics and New Philology. *Computer and the Humanities* 24: 339–343.

Busa, R. 1994. *Inquisitiones Lexicologicae in Indicem Thomisticum*. A Roberto Busa S.I. latino sermone confectae atque a Philip Barras in anglicum sermonem translatae. 2a ed. emendata auctaque. Gallarate: CAEL.

Busa, R. 1998. Concluding a Life's Safari from Punched Cards to World Wide Web. In *The Digital Demotic: a selection of papers from Digital Resources in the Humanities 1997*, ed. Lou Burnard, Marilyn Deegan and Harold Short, 3–11. London: Office for Humanities Communication Publication Number 10.

Busa, R. 1999. Picture a man...Busa Award Lecture. Debrecen, Hungary, 6 July 1998. *Literary and Linguistic Computing* 14(1): 5–9.

Busa, R. 2000. Dal computer agli angeli – 1261 momenti di pensiero [...], 167. Itaca-BVE: Castel Bolognese.

Busa, R. 2003. Quasi a modo di testamento: profezia o utopia?" *Informatica e Scienze Umane – mezzo secolo di Studi e Ricerche* [Strasbourg Eur. Sc. Found. 14–15 Juin 2002], pp. 57–72. Firenze: Olschki.

Ciula, A, Eide, Ø, and Sahle P., ed. 2019. Models and Modelling between Digital and Humanities – A Multidisciplinary Perspective. *Historical Social Research.* Suppl. 31 – Models and Modelling between Digital and Humanities.

Eisenstein, E.L. 1979. *The Printing Press as an Agent of Change: Communications and Cultural Transformations in early Modern Europe, volumes I and II*. Cambridge, England; New York: Cambridge University Press.

Fellbaum, C. 2005. WordNet and Wordnets. In Encyclopedia of Language and Linguistics, Keith Brown et al. eds., 665–670. Oxford: Elsevier. Second Edition.

Flanders, J. 2009. The Productive Unease of 21st-century Digital Scholarship. *Digital Humanities Quarterly* 3(3).

Flanders, J. and Jannidis, F. ed. 2018. *The Shape of Data in Digital Humanities: Modeling Texts and Text-based Resources*. London; New York: Routledge.

Gange, D. 2014. Interdisciplinary Measures: Beyond Disciplinary Histories of Egyptology. In *Histories of Egyptology: Interdisciplinary Measures*, ed. William Carruthers, 64–77. New York; Oxon: Routledge.

Garfield, E. 1955. The preparation of printed indexes by automatic punched-card techniques. *American Documentation* 6(2): 68–76.

Haug, Dag & Jøhndal Marius (2008), Creating a Parallel Treebank of the Old Indo-European Bible Translations. In *Proceedings of the Language Technology for Cultural Heritage Data Workshop (LaTeCH 2008)*, Marrakech, Morocco, ed. Caroline Sporleder, & Kiril Ribarov, 27–34.

Jones, S.E. 2016. *Roberto Busa, S. J., and the Emergence of Humanities Computing: The Priest and the Punched Cards*. New York; Oxon: Routledge.
Jones S.E. 2016b. Tumblr: Roberto Busa S.J. and the Emergence of Humanities Computing. *Roberto Busa S.J. and the emergence of Humanities Computing*.
McCarty, W. 2005. *Humanities Computing*. Hampshire; New York: Palgrave Macmillan.
McCarty, W. 2008. What's Going On? *Literary and Linguistic Computing* 23(3): 253–261.
McCarty, W. 2013. What does Turing have to do with Busa? In *Proceedings of The Third Workshop on Annotation of Corpora for Research in the Humanities (ACRH-3)*, ed. Francesco Mambrini, Marco Passarotti, and Caroline Sporleder, 1–14. Institute of Information and Communication Technologies Bulgarian Academy of Sciences: The Institute of Information and Communication Technologies, Bulgarian Academy of Sciences.
McCarty, W. 2014. Getting there from here. Remembering the Future of Digital Humanities. Roberto Busa Award lecture 2013. *Literary and Linguistic Computing* 29(3): 283–306.
McGann, J. 2004. *Radiant Textuality: Literature after the World Wide Web*. New York; Hampshire: Palgrave Macmillan.
Nyhan, J. and Flinn, A. 2016. *Computation and the Humanities: Towards an Oral History of Digital Humanities*. Cham, Switzerland: Springer.
Passarotti, M. (2014), From Syntax to Semantics. First Steps Towards Tectogrammatical Annotation of Latin. In *Proceedings of the 8th Workshop on Language Technology for Cultural Heritage, Social Sciences, and Humanities (LaTeCH) @ EACL 2014*, ed. Kalliopi Zervanou & Cristina Vertan, 100–109. Gothenburg: The Association for Computational Linguistics.
Pettegree, A. 2011. *The Book in the Renaissance*. New Haven; London: Yale University Press.
Prescott, Andrew. 2012. Consumers, creators or commentators? Problems of audience and mission in the digital humanities. *Arts and Humanities in Higher Education* 11: 61–75.
Ramsay, S. 2011. *Reading Machines: Toward an Algorithmic Criticism*. Urbana; Chicago; Springfield: University of Illinois Press.
Rockwell, G. 2016. The Index Thomisticus as Project. *Theoreti.ca*: *Research notes taken on subjects around multimedia, electronic texts, and computer games*. https://theoreti.ca/?p=6096 (accessed 18 June 2019)
Rouse, M.A., and Rouse, R.H. 1991. *Authentic Witnesses: Approaches to Medieval Texts and Manuscripts*. Notre Dame, Indiana: University of Notre Dame Press.
Sinclair, S. 2016. Experiments with punch cards. *Stefan Sinclair: scribblings and musings of an incorigible digital humanist*. https://stefansinclair.name/punchcard/ (accessed 18 June 2019)
Snow, C.P. 1959. The Rede Lecture, 1959. In C.P. Snow. *The Two Cultures: and a Second Look*, 1–22. Cambridge: Cambridge University Press.
Tasman, P. 1957. Literary Data Processing. *IBM Journal of Research and Development* 1(3): 249–256.
Wisbey, Roy. 1962. Concordance Making by Electronic Computer: Some Experiences with the "Wiener Genesis." *The Modern Language Review* 57(2): 161–172.

Chapter 2 A First Example of Word Index Automatically Compiled and Printed by IBM Punched Card Machines

Roberto Busa S.J.

First published as: Busa, R. 1951. *S. Thomae Aquinatis Hymnorum Ritualium Varia Specimina Concordantiarum. Primo saggio di indici di parole automaticamente composti e stampati da macchine IBM a schede perforate.* Milano: Bocca.

Fig. 2.1. This picture was taken on 27/06/52 in the IBM World Headquarters, New York. Also present, from left to right (as recorded by Busa): G.J. Rebsamen; Rev. Victor Yannitelli S.J.; Prof. William R. Parker; L.P. Pennell; J.G. Phillips; Thomas J. Watson Jr; Charles Mirabile; Paul Tasman; Francis Castelli; Cardinal F. Spellman; Padre Busa; Computer (Busa Archive #0010). Paul Tasman can be seen holding a copy of the *Varia Specimina* (the text reproduced below). He appears to discuss it with Cardinal F. Spellman, as Busa and others look on.

Editors' note

Busa is most celebrated for his work on the *Index Thomisticus*, which was published in hard copy from 1974–80, when concordance making had become routine in the field of Humanities Computing. The text reproduced below was published in 1951, and it is

J. Nyhan and M. Passarotti (eds.), *One Origin of Digital Humanities*,
https://doi.org/10.1007/978-3-030-18313-4_1

thought to be the earliest methodological discussion, and exemplification, of the use of electromechanical accounting machines to partially automate the process of concordance making. In other words, it is the text below, rather than the *Index Thomisticus*, that places Busa at the source of the literary data processing tradition.

When the *Varia Specimina* was being prepared for publication as a bilingual text, Busa's command of English could not yet detect the low quality of the translation that had been provided to him (Roberto Busa to Robert D. Eagleson, July 4, 1966). That the English-language text below can be difficult to follow is due not only to the quality of the translation, it is also due to Busa's conflation of an account of what happened with an account of what would have been possible were certain machines available. Yet, this article presents a fascinating insight into some aspects of Busa's early work. The technologies used in the *Index Thomisticus* would change utterly during the c.30 years of its elaboration. The five stages of concordance making that Busa summarizes here would remain mostly valid (as we shall see, a pre-editing stage will also appear in later publications and lemmatization will be automated).

In this text Busa presents a methodological description of the mechanically-assisted processing of the third canto of Dante's *Inferno.* He also presents concordances of four hymns of Thomas Aquinas. Busa's work was hampered by the limited, 80-space capacity of the punched cards he had at his disposal. Only upper-case text could be represented; accents and punctuation had to be jettisoned. Busa describes how the text was transcribed onto 136 punched cards (by whom he does not say). The cards were then processed by the Reproducer and 943 cards were eventually output, with each word of the canto reproduced on a card along with "its quotation". It was not possible to fully automate this, he notes. After the Sorter had put the cards into alphabetical order the philologist was required. Invariably a 'he' in Busa's writings, the philologist distinguished homographs, and other tricky wordforms, and lemmatized the words of the text. The cards were then alphabetized once more and printed with the tabulator. Busa notes that it would be possible to derive a published edition from a reproduction of the print out. He also discusses other machines like the Collator, which could be used, among other things, to search the cards, though it is difficult to determine which of these machines he actually used at this stage of his work.

Busa closes this article on an optimistic note. He looks forward to using the IBM Cardatype and the IBM 407 accounting machine, which were then available for public use in the United States, but not in Europe. They would allow improved "final printing, punctuation, accents and texts longer than the usual card capacity". As such, the Cardatype would also allow the text that was punched on cards to be represented on tape, thus facilitating the serial processing of his texts. So too, Busa mentions his planned project on Aquinas and his belief that it would be possible to complete it in four years, rather that the half century that might usually be expected. Little could he have guessed that he would spend much of the next thirty years working on that very project.

References

Letter from Roberto Busa to Robert D. Eagleson, July 4, 1966, (Rel. Cult. Australia 1944)

Busa, R. 1974–1980. *Index Thomisticus: Sancti Thomae Aquinatis Operum Omnium Indices et Concordantiae*. Stuttgart: Frommann Holzboog, 56 voll.

I must first state in what sense I am using the three words: *index verborum, concordantia, lexicon,* which form the argument of this paper, for I do not believe that they possess a well-defined meaning for everybody. At all events, they are, especially the first two, unknown to many. For the great majority of men who while passing through the boundless forest of the feasible, have not been allowed to direct their steps down this little sheltered path, I would commence by evoking to them the recollection of that Rhyming Dictionary of the *Divina Commedia* which, an appendix to the Dantesque volume of our youthful years, could be termed not only by the appellative—now traditional—of Improved Rhyming Dictionary, but indeed by that of a spectator to our carelessness in its regard. There we found grouped all the lines ending with the same rhyme. Now, when this is done for each different word, or, when under each word all the lines which contain this word are transcribed, one by one, we have the "concordance". It is well known by the Ministers of the Gospel who have been able to enhance their sermons with abundant biblical quotations, thanks to the friendly pages of one of the many, both old and new, editions of the *Concordantiae Scripturae Sacrae* which can be found in Parish Houses. "Word index" instead, is the term used for a list which gives for each entry the numerical listing of quotations only—canto, for ex[ample], verse and line—without actually repeating the sentences thereof. Both these aids offer a ready field of work for whoever is desirous of preparing the "lexicon" setting forth the various meanings of those words, grouped, classified and illustrated.

It is from the mountains of these indexes, concordances and lexicons, compiled first on the single authors and then little by little collated and amalgamated so as to be lexicons of an epoch or even of a language, it is, as I said, from this mass of works that were derived, or at least should have been derived, those vocabularies which for many years filled the hours and eyes of our boyhood, and of which we would stand so much in need also now, when the serious questions that beset us grownups would fain be expressed in mature, conscious and controlled language.

More substantial observations on the contribution offered by these aids to pure knowledge, to all philology taken in its widest sense, are not only familiar to those competent in such matters, but can also, and easily, be imagined by anyone reflecting on how the structure and history of the spoken word are merely the manifestation of the structure and history of thought. How and what benefits such studies can bring to philosophy and theology, this was actually the determining factor and remains the scope of my investigations. I have outlined elsewhere a number of remarks in this connection.

* * *

Reverting to the nature of the *indices verborum* and *concordantiae* I believe it is easy to realize how much and how hard has been the work required for their compilation. To break up an author's text into phrases, transcribe them on cards, re-transcribe each card as many times as there are words in the phrase, put them into alphabetical order, so as to have in a card file, in the proper order, all, I repeat, materially all the prepositions, conjunctions, adverbs, adjectives, nouns, verbs—for example all the *in*, all the *not*, all the *here,* all the *now*, which have dripped from the author's pen, and so on for each single word of his vocabulary, is obviously equal to copy out seven, twenty and up to eighty times (which is the case of the *Thesaurus Linguae Latinae*) the whole work from the first to the last line, supplementing each card with the addition of the quotation and the order word. It is easily imaginable what it takes to do all this by hand, with nothing but the traditional aid of that same paper, pen and ink which were for so many centuries the only "office machines". Not without reason those who undertook such a task, and naturally persisted in finally reaching its conclusion, always betray their need that others know what it cost them. It was confessed, truly in none too elegant verse, by Charles Conrad Reitz, brother of the perhaps more famous John Frederic, in the epigram with which he prefaces his *Index verborum ac phrasium Luciani* [...]([1]):

> Candide mirari Lector tibi forte subibit,
> fessa sit Indicibus cur mea dextra tribus;
> namque opus Herculeos superans gravitate labores
> auctori misero commoda nulla parat [...]

And perhaps the sole earthly reward he may expect is that it be recognized what price had to be paid the satisfaction of having toiled in order that others might work with greater ease.

Despite this, in 1907 Hermann Schöne([2]) enumerated already some fifty indexes and lexicons of individual Greek authors; in 1914 Paul Rowald([3]) listed 144 for Latin authors, and in 1926 Paul Faider([4]) stated that for the latter there were 39 indexes and 20 lexicons composed after 1850 on critical editions.

The writer also intends to prepare for publication a repertory as complete as possible of the concordances and indexes which have appeared for any kind of writing whatsoever. Meantime, however, I will give here only the names of what has actually passed through my hands. I would point out that in some cases the index or lexicon takes in only some part or aspect of the works and that the town mentioned in the footnote is where the work was published.

Indexes of Latin authors, published prior to 1926: Catullus([5]), Caesar([6]), Cicero([7]), Justin([8]), Lucilius([9]), Lucretius([10]), Minutius([11]), Horatius([12]), Petronius([13]), Plautus([14]), Propertius([15]), Prudentius([16]), Salustius([17]), Svetonius([18]), Tacitus([19]),

---

[1] Utrecht, 1746.

[2] *Repertorium griechischer Woerterzeichnisse und Speziallexika*, Leipzig, 1907.

[3] *Repertorium lateinischer Woerterzeichnisse und Speziallexika, Leipzig*, 1914.

[4] *Repertoire des Index et Lexiques d'Auteurs Latins,* "Les Belles-Lettres", Paris, 1926.

[5] New Haven, Conn., USA, 1912.

[6] Jena, 1886; Berlin, 1837 c 1893.

[7] Jena, 1877–1884 c. 1887–1894.

[8] Hannover, 1882.

[9] Berlin, 1888.

[10] Leipzig, 2 ediz. 1926.

[11] Liege, 1909.

[12] Hannover, 2 ediz. 1879; Washington, 1916.

[13] Leipzig, 1898.

[14] Leipzig, 1904 e 1933.

[15] Oxford, 1905.

[16] Cambridge – Mass., USA, 1932.

[17] Hannover, 4a ediz. 1890.

Virgil([20]); published after 1926: Apuleius([21]), Bede([22]), Boetius([23]), Cicero([24]), Elius Donatus([25]), Lucanus([26]), Ovid([27]), Peregrinatio Aetheriae([28]), Seneca([29]), Servius([30]), Statius([31]), Terence([32]), Tertullian([33]).

Of the Greeks: Andocides([34]), Antiphon([35]), the Apologists (Justin, etc.)([36]), Apostolic Fathers([37]), Aristophanes([38]), Aristoteles([39]), Athanasius([40]), Attic Orators([41]), Demosthenes([42]), Dinarcus([43]), Pseudo Dionysius([44]), Fragments from the Elegiac Poets([45]), Herodotus([46]), Aeschylus([47]), Aeschines([48]), Aesychius of Alexandria([49]), Hesiod([50]), Euripides([51]), Phylodemus([52]), Phylon of Alexandria([53]),

---

[18] Cambridge – Mass., USA, 1922

[19] Lyon, 1900; Leipzig, 1903.

[20] Rouen, 1710; New Haven – Conn., USA, 1911; Leipzig, 1912.

[21] Middletown – Conn., USA, 1934.

[22] Cambridge – Mass., USA, 1929

[23] Camdridge – Mass., USA, 1928.

[24] Haag, 1937; Urbana – Ill., USA, 1938.

[25] Ithaca – N.Y., USA, 1930.

[26] London, 1927; Washington, 1940.

[27] Washington, 1939.

[28] Amsterdam, 1930.

[29] Paris, 1937.

[30] Ithaca – N.Y., USA, 1930.

[31] Washington, 1943.

[32] Chapell Hill – N.C., USA, 1932.

[33] Leipzig, 1904 e 1933.

[34] Oxford, 1897.

[35] Ithaca – N.Y., USA, 1895.

[36] Leipzig, 1912.

[37] Leipzig, 1907.

[38] Oxford, 1883; Cambridge – Mass., USA, 1932.

[39] Berlin, 1870.

[40] Berlin, in pubblicazione dal 1944.

[41] Oxford, 1828.

[42] Leipzig, 1892.

[43] Oxford, 1897.

[44] Louvain, 1941.

[45] Ithaca – N.Y., USA, 1908.

[46] Strasbourg, 1824.

[47] London, 2ª ediz. 1847; Leipzig, 1876.

[48] Leipzig, 1826.

[49] Jena, 2ª ediz. 1867.

[50] Lund (Svezia), 1890.

[51] Leipzig, 1841.

[52] Amsterdam, 1941.

Hysocrates([54]), Lycurgus([55]), Lysias([56]), Lucian([57]), Homer([58]), Pindar([59]), Plato([60]), Xenophon([61]), Sophocles([62]), Theocritus([63]), the Tragic Poets([64]), Tucidides([65]).

Parallel with the development of these particular works runs that of general plans from the *Lexicon Totius Latinitatis* of Forcellini, to the cardfile of the *Thesaurus Linguae Latinae* at Munich, Bavaria, where ten million cards (an enormous "Roman" army at the service of the empire of a civilization, which is no longer either a civilization among others, nor the civilisation of one people among others, but the civilization of all those who on earth are the children of God) record for each word of all the authors, from the primitives right up to Tacitus and Svetonius([66]) a text averaging 80 words (the *et* for example fills 40 drawers of about 1200 cards each); from *the big* in folio [sic] of the *Thesaurus Graecae Linguae*, 1571, of Henry Estienne (the II), to the present *Archiv für Griechische Lexikographie* of Hamburg; from the *Glossarium ad scriptores mediae et infimae latinitatis* and the *Glossarium ad scriptores mediae et infimae graecitatis* of Du Cange (respectively 1678 and 1688) to the Dictionary of Mediaeval Latin, today resumed by the *Union Académique Internationale* in Paris, and by the *Mittellateinisches Wörterbuch* at Munich, Bavaria.

Still within the field of Letters but outside that of classic and post-classic Latin and Greek authors, I was able to handle the concordances of the Italians Dante([67]), Petrarch([68]); of the German Walter von der Vogelweide([69]), of the English *Beowolf*([70]), R. Browning([71]), R. Burns([72]), Chaucer([73]), S.T. Coleridge([74]), J.

---

[53] Berlin, 1926–1930.

[54] Leipzig, 1904.

[55] Oxford, 1897.

[56] Bonn, 1895.

[57] Utrecht, 1746.

[58] Oxford, 1880; Leipzig, 1885 e 1891; Leyden, 1905; Halle, 1914.

[59] Leipzig, 1883.

[60] Leyden, 1789; Oxford, 1832; Leipzig, 1835–1836.

[61] Leipzig, 1801–1802; Leipzig, 1869; Ithaca – N.Y., USA, 1900.

[62] Berlin, 2ª ediz. 1872; Leipzig, 1870.

[63] Leipzig, 1879.

[64] Petropolis, 1892.

[65] Genève, 1843 e 1847; Berlin, 1887.

[66] For later authors not every word has been filed on a card but only selected words.

[67] *Divina Commedia*: Cambridge – Mass., USA, 1888; *Opere Italiane*: Oxford, 1905; Cambridge – Mass., USA, 1936; *Opere Latine*: Oxford, 1912.

[68] Oxford, 1912.

[69] The University of Wisconsin – USA, 2ª ediz. 1950.

[70] Halle, 1911.

[71] New York, 1924–1925.

[72] Glasgow, 1889.

Donne([75]), Thomas Gray([76]), Robert Herrick([77]), Thomas Kyd([78]), Marlowe([79]), Milton([80]), Shakespeare([81]), Tennyson([82]), W. Wordsworth([83]), T. Wyatt([84]), authors of Middle English([85]); the Americans R.W. Emerson([86]), S. Lanier([87]), E.A. Poe([88]). For the Law: Theodosian's *Index*([89]), the Constitution of the USA([90]), the *Codex Juris Canonici*([91]). For religious literature: *The Christian Year*([92]), *S. Teresa de Jesús*([93]), *San Juán de la Cruz*([94]), *De Imitatione Christi*([95]), post-biblical Hebrew literature([96]), the *Rules of St. Benedict*([97]), and may I be allowed to add the names of the concordances of the *Koran*([98]), and of Luther's writings([99]), although l did not actually handle them.

I have not taken a direct interest in biblical concordances. The most important specialized encyclopaedias give a complete information about them, accessible to all. Historically speaking it seems they were the first([100]). Among the Latin concordances (I am speaking of verbal ones, not of the so-called *concordantiae reales*, or analytical indexes) there has been from the first, terminated in 1230, up to

---

[73] Washington, 1927.

[74] Saint Mary of the Woods – Ind., USA, 1940.

[75] Chicago, 1940.

[76] Boston and New York, 1908.

[77] New York, 1936.

[78] Louvain, 1906.

[79] Louvain, 1911–1932.

[80] Madras, 1857; London, 1867; London, 1894; Halle, 1923.

[81] Philadelphia, 1902; London, 1937; New York, 1937.

[82] London, 1869; New York, 1914.

[83] London, 1911.

[84] Chicago, 1941.

[85] New York, 1943.

[86] New York, 1932.

[87] Austin – Tex., USA, 1939.

[88] Baltimore, 1941.

[89] Berlin, 1925 e 1929.

[90] New York, 1872.

[91] Città del Vaticano, 1941.

[92] Oxford, 1871.

[93] Burgos, 1945.

[94] Burgos, 1948.

[95] London and New York, 1911.

[96] Mishnah: Jerusalem, e Frankfurt a. M., 1914–1927; Thosephta: Jerusalem, 1932–1940. Targum Onqelos: Jerusalem, 1932–1940.

[97] Klagenfurt, 1692; Augsburg, 1723; Styrae, 1763; Maredsous, 1946.

[98] Leipzig, 1842.

[99] Darmstadt, 1827–1829.

[100] ...Probably biblical concordances were already in existence in the VI–VII century A.D.

the present day a crescendo of effort to attain a fuller completeness which is deeply significative of the exigencies of these aids (first in 1433 at the Basel Concilium and two years later as its delegate to Constantinople, the Domenican John Stoikowic from Ragusa had found himself suddenly needing, just in the very depths of high theological controversies, the concordances of *nisi,* of *ex*, of *per*...); in 1496 *concordantiae majores*; in 1555 *revera majores*; *quales nemo hactenusest aggressus* in 1562; *maximae* in 1568; up to the *maximae et absolutissimae* in 1600, the *absolutissimae* of 1627, so it was inevitable that manual summaries should be educed [sic] from them. From the Latin the concordances passed to the first Hebrew ones in Venice in 1523, the first Greek by N.T. at Basel in 1546, and F.T. at Frankfurt in 1607, to the Leyden Syriac concordance in 1709, the first German in 1524, English in 1540, Dutch in 1615, Swedish in 1734, Danish in 1748 and French in 1840.

The lexical blossoming of Germany in the last century extended itself to England and has made considerable developments in the United States of America: the United States have, indeed, given us, in the lately deceased Prof. W.A. Oldfather, a master of lexicographical method just as Germany gave us Wöllflin; they have also given us in the *Lexicon Plautinum* (1904 and 1933) by Prof. Gonzales Lodge a jewel which unites within its pages all the characteristics of the index, the concordance and the lexicon, the virtues of a conscientious and intelligent elaboration: and in Prof. R.J. Deferrari, still living in Washington, the organizer of the great concordances of Ovid, Statius, Lucanus and Prudentius. All this serves to confirm the evaluation and therewith the great services which works of this kind render, as I was saying before, to knowledge; and the more so when, as we mentioned, due account is taken of the exhausting amount of work entailed and the sheer materiality of a great portion of it.

This latter aspect strikes everyone, even the least expert, and it is sometimes taken as a reason for despising such work and bestowing compassion upon those who "waste" energy which might have been utilized better elsewhere; but only those who have had a hand, or at least an eye, in this sort of work will realize that it is not a matter of purely material labour. To organize an unbounded mass of material, in function to definite services to be rendered to exigent disciplines, requires intelligence, training to foresee and predispose, caution and vigour so as not to let oneself be induced to take those unwise steps, sometimes shortcuts sometimes bypaths, which occasionally end by frustrating years of work; these last experiences which, being for the most part unpublished, are ignored by many and soon forgotten by all, would be useful, if known, to many.

Nor can such work be successfully concluded unless one is competent or at least possess a well-grounded introduction to the reflective study of thought and expression, knowledge of the grammatical and metaphysical structure of the language, in short, philology with a smattering of philosophy. So much is to be said in defence of the "scientific" nature of such works which, as instruments for further research, are related to the other studies like the obscure buried foundations of a building are related to the visible façade and the splendour of the summit. The ancient fable of Menenius Agrippa rings true also here: to appreciate as science only the splendour of a well admired point which overlooks everything else, would be to consider science not as an edifice to be built for the good of the human family, but as a pedestal for a monument of glory.

These considerations do not, however, prevent our being able, and indeed obliged, to think of aids which make such studies quicker and easier. It is with this that I am dealing here.

I bring down to five stages the most material part of compiling a concordance:

1. transcription of the text, broken down into phrases, on to separate cards;
2. multiplication of the cards (as many as there are words on each);
3. indicating on each card the respective entry (lemma);
4. the selection and placing in alphabetical order of all the cards according to the lemma and its purely material quality;
5. finally, once that formal elaboration of the alphabetical order of the words which only an expert's intelligence can perform, has been done, the typographical composition of the pages to be published.

A kind of mechanization has been working for years so far as regards caption 2: the T.L.L. and the *Mittellateinisches Wörterbuch* use the services of Copying Bureaux, where one of the many well-known systems of duplicating are used; Prof. J. Deferrari of Washington used electrical typewriters which can make many copies; Prof.P. O'Reilly of Notre Dame (Indiana—USA) had each side of the page repeated as many times as there were words contained thereon. But something more comprehensive had to be found which would if possible meet all the requisites of the problem.

Of the systems I have examined up to the present time, two appeared to assure such a response: the *Rapid Selector* and the punched card electrical accounting machines. The former operates on micro-film, which carries, beside the photogram of the page, another, parallel one, bearing the symbols—coded in combinations of white and black spaces—of the words of that page. Before the last war Dr. Vannevar Bush of the Massachusetts Institute of Technology outlined this machine but it was completed by Mr. Ralph Shaw, librarian of the Department of Agriculture, Washington, where I was able to see it operating already in November 1949. Its principal feature is the whirlwind speed with which it explores the reels of microfilm—10,000 photograms per minute—and instantaneously rephotographed on another microfilm strip all and only those photograms which bear a determined item([101]).

I shall not give a detailed description because I thought not suitable to apply this system, to the composition of concordances; I will only say that, besides not allowing automatic printing of the concordances, such as can be done with the sys-

---

[101] Cfr. anche [see also] John C. Green: The rapid selector – An automatic library, in *Rev. Doc.*, XVII (1950) fasc. 3; pg. 66–68; Carl S. Wise and James W. Perry: Multiple Coding and the Rapid Selector, in *American Documentation*, I, n. 2 (Spring, 1950), pg. 76–83.

tem hereunder, the rapid selector necessitates on the one hand that all the cards, to be made from the sorted microfilm, be of photosensitive paper, and on the other hand all the different words and forms of each word be previously coded, for the entire text must be translated into numerical symbols by hand.

Now what I intend publishing, are the results of a first series of experiments carried out with electric accounting machines operating by means of punched cards. Of the three companies using this system, the International Business Machines (IBM), the Powers of the Remington Rand, and the Bull, it was at the Milan Head Office of the Italian organization of the first, which is also the most important, that I continued the research I had commenced at the New York Headquarters. What had at first appeared as merely intuition, can today be presented as an acquired fact: the punched card machines carry out all the material part of the work mentioned under captions 2, 3, 4 and 5.

I must say that if this success has its origin in the multiple adaptability, characteristic of the equipment in question, it was nonetheless due to the open mindedness and intelligence of the IBM people, who have honoured me with their patient confidence, that the method for such application has been found. I will give a brief description of the stages of the process and the first trials which were carried out on one of Dante's Cantos.

The *Automatic Punch*, controlled by a keyboard similar to that of an ordinary typewriter, "wrote" by holes or perforations, one for each card, all the lines: a total of 136 cards. This is the sole work done by human eyes and fingers directly and responsibly; if at this point oversights occur, the error will be repeated from stage to stage: but if no mistakes were made, or were eliminated, there is no fear of fresh errors: human work from now onwards is reduced to mere supervision on the proper functioning of the various machines.

The contents of each card can be made legible either on the punch itself which, if required, can simultaneously write in letters on the upper edge of the card what is "written" in holes on the various lines of columns thereon; or else on a second machine, the so-called *Interpreter*, which transcribes in letters the holes it encounters on the cards (previously punched). This offers not only a more accurate transcription in virtue of the better type and greater spacing of the characters, but a transcription which can be effected [sic] on any desired portion of the card.

The 136 cards thus punched were then processed through a third machine, the *Reproducer*: this automatically copied them on another 136 cards, but adding, sideways of the lines and their quotations, the first of the words contained in each. Subsequently it makes a second copy, adding on the side the second word, then a third copy adding the third, and so forth. There were finally 943 cards, as many as were the words of the third canto of Dante's Inferno; thus, each word in that canto had its card, accompanied by the text (or rather, here, by the line) and by the quotation. This is equivalent to state that each line was multiplied as many times as the words it contained. I must confess that in actual practice this was not so simple as I endeavoured to make it in the description: the second and the successive words did not actually commence in the same column on all cards. In fact, it was this lack of determined fields which constituted the greatest hindrance in transposing the system from the commercial and statistical uses to the sorting of words from a literary text. The result was attained by exploring the cards, column by column, in order to identify by the non-punched columns, the end of the previous word and the commencement of the following one; thus, operating with the *Sorter* and *Reproducer* together, were reproduced only those words commencing and finishing in the same columns.

This operation is rather a long one: theoretically as many sortings and groups of reproductions as there are columns occupied by the longest line, multiplied by the number of letters contained in the longest word; in practice various devices make it possible to shorten this routine a good deal. It must be borne in mind that the amount of human work entailed by all this processing in the *Sorter* and setting up of the reproducer panels—about two persons' one day work—remains unchanged notwithstanding the increased number of cards. While it is true that there are longer intervals, namely those intervals during which the machines carry out their own operations, it is equally true that the operations which in the case of few cards are inevitably consecutive, with many cards can be simultaneous: the time taken by the *Reproducer* to copy one stack can be used to sort others or to set up the panel for the next reproduction. At present the *Reproducer* can reproduce 6,000 cards an hour, and the *Sorter* can explore 36,000.

Having reached this point, it is a trifle to put the words into alphabetical order: the *Sorter*, proceeding backwards, from the last letter, sorts and groups gradually, column by column, all the identical letters; in a few minutes the words are aligned and the card file, in alphabetical order, is already compiled. This order can be obtained again with the same ease, as often as required. If the scholar, while making his research on the varied conceptual content, disturbed the alphabetical order of the items, this same order can be very easily obtained once more merely by the use of the *Sorter*, which is the most elementary IBM machine.

The philologist, however, must group or sort further on what the machine has not been able to "feel"; thus *have*, *had*, are different forms of the same verb; thus, in Italian, *andiamocene*, *diamogliene* are several words joined into one; and for the Latin *mortuus est* is a single word form which means died, but could also mean the dead man is and then they would be two items; and so on for the whole wide range of homonyms. When the order has thus been properly modified and attains its final form, the cards are ready to be processed in the *Alphanumerical Accounting Machine*, or *Tabulator*.

The tabulator retranscribes on a sheet of paper, in letters and numbers—no longer in holes—line after line, the contents represented by the holes in the cards, at the rate of 4,800 cards per hour; and this is a page of the concordance or index in its final arrangement. The published edition can now be obtained by some kind of reproduction; for ex. employing ribbon and paper of the kind that allows the use of lithographical duplicators. The concordance which I am presenting as an example is precisely an off-set, reproduction of tabulated sheets turned out by the accounting machine. The flexibility of these machines offers the possibility of making varied and sometimes extremely useful, applications. I am making a brief mention of the most salient ones. The tabulated document can be printed on a continuous paper roll or else on separate sheets of varying sizes; in other words, the machine can be made to change the sheet automatically after a given number of lines.

The distance between lines can also be automatically differentiated; it is possible to arrange the machine so as to make, for example, without further human intervention, a double space when it goes on to a new word (for example from *anima* to *animato*) and, say, four spaces between the words commencing with the letter A and those commencing with B, and so on. The data which are, for example, at the right of the card can be tabulated, if desired, at the left, vice versa; so that the quotation can be placed prior or subsequent to the line independently of its position on the card.

The card contents can be reproduced also partially, which makes it possible to obtain only an index of the quotations for those words of which it is not deemed desirable to have the concordance. The tabulator's performance is extremely useful when, to use the current technical phrase, it is running in tab. Then it turns out only the list of the words which are different: if, for example, the cards containing the preposition ab total two hundred, the machine will print ab once only, but, if desired, will add at the side thereof the number of times, that is 200, and so on for each word. The list thus obtained is very useful in studying those intelligent integrating touches to be given to the alphabetical order of the words, which, as I said, is effected [sic] by the machine on the mere basis of the purely material quality of the printed word. It is also useful as an entry table for all who wish to peruse the whole vocabulary of an author for determined purposes; still more useful when beside the word it is shown the frequency with which it is used. When another machine called the *Summary Punch* is connected to the accounting machine running in tab, while the latter is turning out the long tabulated list of different words, the former, electrically controlled by the accounting machine, simultaneously punches a new card for each of these words, thus providing ready headings to be placed before the single groups of lines or quotations. If necessary, these can be inserted in their proper place among all the others automatically by the *Collator*.

This *Collator* which searches simultaneously two separate groups of cards at the rate of 20,000 per hour, and can insert, substitute and change cards from one with cards from the other group, also offers some initial solutions to the problem, of finding phrases or compound expressions. Taking, for example, the expression *according to*: the group of cards containing *according* and that containing *to* are processed in the machine; on the basis of the identical quotation, the machine will extract all those cards on which both appear. It is true that they may be separated by other words, but one thing is certain, namely that all the cards bearing *according to* will be among those extracted; the eye and the hand must do the rest. It is still easier to obtain the same result when a card hearing the phrase sought for can be used as [a] pilot-card.

The *Collator* can also be used to verify and correct the cards which have been manually punched at the beginning, and thus guarantee the accuracy of the transcription, an indispensable condition for philological works, particularly in the light of their peculiar function. Two separate typists punch the same text, each on his own; the *Collator* compares the two series of cards, perceiving the discrepancies: of the cards not coinciding, at least one is wrong. This control allows only the following case to pass unobserved, namely two typists make the same error in the same place. This case is very improbable and so much the less probable inasmuch as the qualities and circumstances of typing and typist are different.

This method of verifying, although substantially the same, offers perhaps some advantages over the other, usually employed by IBM in the intent of not doubling the number, and consequently the cost, of the cards purposely, whereas in our case this is no hindrance, since each card already has to be multiplied as many times as the words it contains; the punched cards are put through the *Verifier* on the keys of which a typist repeats the same text; the machine signals him when his punching does not concord with the existing holes; one of the two is wrong.

* * *

Before concluding, a criticism, of these initial results should be made, also to justify the lines along which I am working to perfect the method: only the first man happened to begin his life as an adult.

In the first place, the machines I used—those commonly used in Europe up to 1950—produce a final tabulated page the appearance of which is still perceptibly less satisfactory than that of printed material. Many will hold the opinion that this is compensated by the automatic performance and the high speed of their writing. But it is indeed hard to sacrifice accents and punctuation as well as the difference between capitals and small letters. Similar considerable limits are involved by the card capacity: eighty spaces.

Since each card includes both quotation and lemma, the average text for each word could not therefore surpass, by much, a hendecasyllable. And this is little, the more so if one bears in mind that the machines do not allow the omission of subordinate phrases or even words, by which the pen worker instead can choose only those few words which constitute the substance of an expression. This brevity in the text, perceptible in a printed concordance and even more so in the case of prose instead of verse, is extremely distressing when the card file is used for research work; infinite occasions will indeed arise where the scant surrounding will not give the lexicographer sufficient elements for a well-grounded interpretation and, by compelling him to a too frequent and aggravating recourse to the text, will tempt him—there are even little devils specialized in leading philologists into sin! —with the bait of a hasty judgment.

Even with only the groups of machines above mentioned, it is quite possible to obviate this latter hindrance, but I will not set forth the various means of doing this. Not only so as not to disconcert the reader: it does happen indeed that when one glimpses at the unimagined possibility of carrying out, for example, in four years a work which would have required otherwise half a century (this is the case of the concordance I have in mind for 13,000 in *folio* pages of the Works of St. Thomas Aquinas) everyone becomes so confident and at the same time so exacting with the new method, that all feels deluded when told that the operations involved in making it possible to have an abundant text on every card will delay, let us say, by twelve months, the conclusion of the work. But it would above all be purposeless to devote time and attention to such devices, for new model IBM machines already in public use in the United States, but not yet in Europe, will allow a more aesthetically precise final printing, punctuation, accents and texts longer than the usual card capacity. I refer to the *Cardatype* and the type *407 Accounting Machine*. I hope to write about this in the near future.

* * *

It is at all events certain that from now on the history of concordances will no longer have to record figures like those of the past: five-hundred Domenicans [sic] —can it really be true?—employed by Hugh de Saint Cher in 1200 in Paris for the first biblical Latin concordance; fifty monks occupied in preparing the biblical concordances organised by the Benedictines in 1700; five German universities cooperated to set in order by hand the ten million cards of the T.L.L. at the end of the last century.

Today it will suffice that man's hands transcribe by typing the entire text on the Punch, control its accuracy either perusing the cards which the Interpreter will have made legible or using, as I have mentioned, the *Verifier* or the *Collator*; this done, the *Reproducer*, the *Sorter* and the *Accounting Machine* and *Collator*([102]) will take entire care of the remaining material part of the work; in a few days the philologist, with the mere help of a technical expert for the care of the machines, will have in hand the general card file and the final proofs corrected for the printer, certain of an accuracy which could never have been guaranteed by the cooperation of man's sensorial and psychical nerve centres.

Another occasion will be taken to illustrate various interesting extensions of this system regarding the most diverse purposes. For example, publication of any kind of repertories, of *incipitari* for Codexes, permutation tables for multilingual dictionaries—in a word, wherever the same group of written signs—numbers, symbols, words or sentences, it makes no difference—has to be put into order and sorted from different and reciprocal aspects. From a strictly philological viewpoint; it will be possible to make extremely quick comparative analyses of the composition and frequency of the vocabulary of various authors, useful for example in psychological research, criticism of texts or historical relationships; searching for all the words containing a given root; glottological comparisons of differ-

[102] Ecco i corrispondenti nomi in inglese [Here are the corresponding names in English: perforatrice: punching machine; verificatrice: verifier; interprete: interpreter; interprete: interpreter; riproduttrice: reproducer; selezionatrice: sorter; tabulatrice: accounting machine; inseritrice: collator; perforatrice riepilogativa: summary punch].

ent languages and so forth. It is by these systems that the science of documentation is endeavouring to reach that level which will enable it to make a timely arrival at that point from where comes its efficiency today more than ever demanded by the developments of study and work. And who knows that these may not be, at least for some decades, the new methods called for by *bibliotechnique* everywhere, dismayed by the technical and financial problems involved in marshalling the unceasing downpour of publications into boundless but flexible card files.

II

For the text of this first example of mechanically compiled concordances I have chosen some pages from St Thomas Aquinas because it is my intention to complete the general alphabetical card file of all his works([103]), thus facilitating the rapid publication of its concordances, to the extent and in the manner, which will be deemed appropriate.

Among the writings I have given preference to the Hymns from the opuscule *Officium de festo Corporis Christi*, because the quantity of those lines came effortlessly within the limits imposed, at least for these first trials where some uncertainty was felt, by the eighty columns of the card. It did not seem advisable to add to these hymns the well-known *Adoro Te Devote*, the Thomistic authenticity of which is much more challenged([104]). I am therefore dealing with four poetic compositions by St. Thomas Aquinas: *Pange lingua, Sacris solemniis, Verbum supernum*, taken respectively from the Vespers, Matins, Lauds of the Office, and *Lauda Sion*, which is the sequence of the Mass.

I shall first give the text, reprinted on the accounting machine from punched cards: in the numeration the line number follows that of the verse, but does not depend on it, because it continues consecutively throughout the whole hymn. I do not follow any critical edition, because there are none as yet, nor had I any intention of making any textual criticism myself([105]). I have adopted the Parma edi-

---

[103] *Cfr.* [See] Complete Index Verborum of Works of St. Thomas in *Speculum* (The Mediaeval Academy of America), XXV, 3 (1950, July), pg. 424–425.

[104] Recent reserves as to authenticity of the oposcule *Officium de festo Corporis Christi* have been renewed this year by H. Vanderhove.

[105] Varianti di qualche rilievo appaiono dal semplice raffronto delle piú comuni edizioni. Elenco quelle che risultarono dal confronto di quattro edizioni del l'Opera Omnia (Piana del 1570, Veneta prima del 1593, Veneta altera del 1754, Parmense del 1850), delle edizioni degli opuscoli di Venezia del 1587, di P. Mandonnet a Parigi nel 1927, e dell'uso liturgico odierno (Messale e Breviario romani):

(variants, of some importance, appear from a simple comparison of the most common editions):

1) P 4,21: *Fit* sanguis Christi merum

Opera Omnia	1570: fitque
	1593: fitque

	1754: fit
Opuscoli	1587: fitque
	1927: fitque
Liturgia	: fitque.

2) S 5,17: Sic sacrificium *istud* instituit

Opera Omnia	1570: [ ]
	1593: [ ]
	1754: istud
Opuscoli	1587: [ ]
	1927: istud
Liturgia	: istud.

3) S 6,22: Dat panis *caelicus* figuris terminum

Opera Omnia	1570 : caelicus
	1593: caelitus
	1754: caelitus
Opuscoli	1587: caelicus
	1927: caelicus
Liturgia	: caelicus.

4) S 7,26: Sic nos tu *visita* sicut te colimus

Opera Omnia	1570: visitas
	1593: visitas
	1754: visita
Opuscoli	1587: visitas
	1927: visita
Liturgia	: visita

5) L 6,17: In qua mensae *primae* recolitur

Opera Omnia	1570: prima
	1593: prima
	1754: primae
Opuscoli	1587: prima
	1927: prima
Liturgia	: prima.

6) L 10,28: *Docti sacri* institutis

Opera Omnia	1570: Docti sacris
	1593: Docti sacris
	1754: Docti sacris
Opuscoli	1587: Doctis sacris
	1927: Docti sacris
Liturgia	: Docti sacris.

7) L 15,45: Integer *assumitur*

Opera Omnia	1570: assumitur
	1593: assumitur
	1754: assumitur

tion's text(106) which is both the most diffused and, after the splendid new edition issued in recent years in the United States(107) the latest of the complete editions of the Opera Omnia. I have modified but once the Panna text on P 4,21: *Fit sanguis Christi merum* to which I have preferred the reading *Fitque sanguis Christi merum*, which seemed to me obviously called for by this hymn's metrical rhythm [sic], where the uneven lines are octosyllabic and the even ones heptosyllabic: on the other hand, this would be the sole exception to the regularity of the first three rythmical compositions, the structure of which, after all, is so rudimental; whereas the dissymetries in the *Lauda Sion* were—I believe—granted St. Thomas by the very nature of a sequence. Furthermore, I have omitted the *Amen* which closes every hymn; and unfortunately, it was only when the indexes were already printed that I noticed that lesson L 10,28: *Docti sacri institutis* is, if the meaning is considered, so evidently unsustainable as to be taken for a mere printer's error: therefore the word sacri at No. 319 of the *Laterculum primum* and at No. 242 of the *Rationarium*, should be deleted from the Thomistic diction of these pages and substituted by *sacris*.

Then we come to the first *Laterculum verborum*. This is the list of all the different words recurring in the text, materially catalogued according to the crudeness of the written word just as they come from the accounting machine running *in tab*. Consequently this list considers *quam* and *sui* each as a single word, while actually the first is an adverb and the accusative feminine singular also of the relative pronoun *qui quae quod*, and the second is both dative of the pronoun *sui sibi se* and nominative plural masculine of the adjective *suus sua suum*; and again, *fit* is considered different from *fitque*, which is here treated as one single word, whereas it is the verb fit plus the enclitic *-que*; the same applies to *genitoque, sanguinisque, trinoque, unaque*. These observations gave rise to the distinction and the integration between the first *Laterculum verborum* and the *Rationarium verborum*.

This also elucidates the fact that the first index enumerates 447 words, while the second has 450, just on account of the words *genito, trino, una*, inserted in numbers 239, 234, 30 of the *Index a contrario*, which, contrary to *fit* and *sanguinis*, are never found free of the enclitic *-que*. In this first index each word is pre-

---

Opuscoli	1587: assumitur
	1927: accipitur
Liturgia	: accipitur.

8) After L 16,48 the Lithurgie and Mandonnet insert the following verse which is lacking in other editions:

Sumunt boni, sumunt mali,
Sorte tamen inaequali
Vitae vel interitus

106 Tomo XV, Parma 1864, Opusc. V, pg. 233–238.

107 *Editio prima Americana*, Musurgia, New York, 1948–1950.

ceded by its progressive number and followed by the indication first of the frequency and then of the number under which it is grouped in the *Rationarinum*.

The second *Laterculum verborum, contrario litterarum ordine ab ultima ad primam dispositum*, groups the same words in reverse alphabetical order, that is commencing from the last letter: it is the *Konträrindex* of the Germans, the *Reverse Index* of the English. The purpose of such aids is easily comprehensible; they are useful on one hand in reconstructing texts—damaged tomb or memorial tablets, papyri, parchments, folios...—where mutilated words have to be reformed and on the other hand in researches concerning desinences and terminations of words. I have taken as my model the big Rhyming Dictionary of the Italian Rosasco([108]), the works of Gradenwitz([109]), and of Buck([110]). Here, too, each word is preceded by the progressive number belonging to this list and followed by the indication of the frequency.

The third index *Rationarium verborum* groups under the respective lemma or identity word the different forms of the same word. This is the first elaboration where the scholar has to use his own competency and discernment and it is the fundamental guide for ordering the *Index verborum* and *Concordantiae*. The identity words or lemmas—I have followed Forcellini in formulating them—have a numeration of their own. Obviously, I do not consider these headings as part of the text, but under each I had the machine list the forms and the number of the times in which that word is used, even if they coincide with the forms appearing in the heading: therefore, under its own lemma number each word carries with it the number under which it is aligned in the first *Laterculum*. I have had the various forms of the same word listed, for obvious reasons, in alphabetical order, not in the grammatical one used for paradigmas of declinations and conjugations.

The *Conspectus lemmatum rationarii verborum* offers the facility of perusing a dictionary of the author in question, classified and summarized by the single entries. The fifth index is the *Index verborum*, in the particular and precise meaning referred to at the beginning of this paper: it lists for each word only the numerical quotation of every excerpt in which it recurs in any of its forms. The *Index verborum* is therefore nothing else but the *Rationarium*, with the sole difference that, instead of the number indicating the frequency, the machine has this time transcribed all and only the single quotations. The machine, however, could equally have transcribed and set in order everything according to the consecutive order of the quotations. I have arranged also for each word to be repeated every time, but I could have asked the machine to transcribe only the different words and to list for each of them the various quotations. Take, for example, [Table 2.2]:

---

108 *Rimario* di Girolamo Rosasco. Padova, 2ª ediz. 1819.

109 *Laterculi vocum latinarum – voces Latinas a fronte et a tergo ordinandas* curavit Otto Gradenwitz, Leipzig, 1904. *Heidelberger Konträrindex der griechischen Papyrus-urkunden* – Leitung: Otto Gradenwitz [ecc.]. Berlin, 1931.

110 *A reverse index of Greek nouns and adjectives* [...] by C.D. Buck and W. Petersen. The University of Chicago Press, (194...) [sic].

SUM ES ESSE					SUM ES ESSE			
ESSE	L	18	54		ESSE	L	18	54
EST	L	17	49		EST	L	17	49
SINT	S	1	1		SINT	S	1	1
SINT	S	1	3			S	1	3
SIT	L	5	13		SIT	L	5	13
SIT	L	5	13			L	5	13
SIT	L	5	14			L	5	14
SIT	L	5	14			L	5	14
SIT	L	17	51			L	17	51
SIT	P	6	34			P	6	34
SIT	P	6	36			P	6	36
SIT	V	6	22			V	6	22

Table 2.2. *Sum Es Esse*

Finally, the *Concordantiae verborum*. This index, too, is but the *Rationarium*, with the difference that the whole transcription of the single lines, and no longer—as for the *Index verborum*—of the quotation only, replaces the brief numerical indication of the frequency. The machine has been set to place the lines in the numerical order of the quotations.

The 165 lines of the four hymns in the volume are thus seen to contain a total of 614 words, grouped in 450 different forms of 335 different words. The six indexes into which they were arranged were compiled in a few hours by means of IBM machines. Human work consisted in transcribing the four hymns only once on to the cards by typing on the punch, in studying the first index which the machines turned out, that is the first *Laterculum*, to determine the identity words or lemmas; in having these lemmas punched and finally in placing under each the words pertaining to them; although if desired this last work, too, could be done by the machines.

* * *

I have in this way given an example, which I believe it is an alluring one, for the great and small facilities with which accounting and statistical machines can bestow index material to men of letters. Other facilities can be excogitated. Anyone wishing to inform us of what may appear to him or faulty to be eliminated [sic] or desirable to be added, will, besides having given us a coveted recognition of our effort's utility, contribute also to a more proficuous [sic] work for many others.

Robert Busa, S.J.

From the Aloisianum at Gallarate, August 15, 1951.

Chapter 3 The Use of Punched Cards in Linguistic Analysis

Roberto Busa S.J.

First published as: Busa, R. 1958. The Use of Punched Cards in Linguistic Analysis. In *Punched Cards – Their Application to Science and Industry,* 2nd edition, ed. R. S. Casey, James W. Perry, Madeline M. Berry and Allen Kent, 357–373. New York: Reinhold.

Fig. 3.1. This photo was taken the year that the article below was published. It shows Busa presenting the *Index Thomisticus* project at Expo '58, the 1958 World's Fair in Brussels. 03/09/58 (Busa Archive #0127)

Editors' note

By the time of the writing of this article, the IBM 407 accounting machine, operating with an IBM 26 card punch, had become available to Busa. Here he describes how these and other accounting machines were used by his project to represent and process texts, line by line, and then word by word, on punched cards. He also describes tools like frequency lists, concordances and indexes that could be printed as a result. Busa emphasizes the usefulness of punched cards for the analysis of authors other than Aquinas, and for languages other than Latin, including right to left writing systems like Arabic and Hebrew. He also discusses the technical constraints on his work (for example, with

J. Nyhan and M. Passarotti (eds.), *One Origin of Digital Humanities*,
https://doi.org/10.1007/978-3-030-18313-4_2

respect to card capacity and character coding) and comments on financial and staff-time related issues.

In this article Busa outlines the pre-editing stage, which involved the analysis and markup of text in preparation for its encoding onto punched cards. This stage had been omitted from discussion in the *Varia Specimina* (see Chap. 2). He also indicates that manual intervention remained a crucial part of the workflow of the *Index Thomisticus* (for example, the annotation of punched cards with mark sensing technology). As such, what he describes is an integrated process of automatic and manual interventions to prepare and process the data. Regarding his work on the *Index Thomisticus*, he portrays this, and the work of mechanically assisted literary analysis, as being heavily constrained by the capabilities of the technology he had at his disposal. The picture is very much one of Busa attempting to coax accounting machines in directions that their makers had not necessarily envisaged by augmenting their processes with human labour.

A chapter describing the application of punched cards to studies in the humanities has a place in a book such as this, because such application in part parallels, and in part coincides fully with its corresponding use in scientific documentation and in libraries. I maintain that it is mutually advantageous to consider how the same tool for investigation responds to the demands of many problems differing in their nature.

I am concerned here with "linguistic analysis" in a broad sense, rather than in any of the specific meanings that different schools have sought to impose upon the phrase. I refer to any type of investigation of language, whatever the significance the word "language" can assume. For example, I include the study of phonetics, of glottology, grammar, or style. In a word, I speak of philology in its broadest sense, and of psychology. I speak only of the investigation of written material, or more strictly, printed words. Even studies of phonetics can be based on printed texts. Hence, I am not concerned with those other analyses dealing directly with human sounds, such as those conducted at the Haskins Laboratories[1] in New York, nor with devices on which data are not recorded in letters or dies of comparative phonetics.

The analysis of language is as old as the knowledge of human knowledge. Even without disturbing Plato in his Dialogues, it would be necessary only to recall to mind the rhyming dictionaries and the hundreds of concordances that have been

---

[1] These studies are aimed primarily at isolating significant signals embedded in the speech stream and in analysing their perception by the human listener. An analysis-synthesis technique is used wherein the speech is converted into visible patterns, the patterns are re-drawn in simplified form and, finally, the modified patterns are re-converted into synthetic speech to provide the acoustic stimuli for perceptual studies.

"Some Results of Research on Speech Perception," A. M. Libermann, *The J. Acoust. Soc. Amer.*, 29, No. 1, 117–123 (1957).

"The Interconversion of Audible and Visible Patterns as a Basis for Research in the Perception of Speech," F. S. Cooper, A. M. Libermann, and J. M. Borst, *Proc. Natl. Acad. Sci.*, 37, No. 5, 318–325 (1951).

"Some Experiments on the Perception of Synthetic Speech Sounds," F. S. Cooper, P. C. Delattre, A. M. Libermann, J. M. Borst, and L. J. Gerstman, *J. Acoust. Soc. Amer.*, 24, No. 6, 597–606 (1952).

published since the invention of printing. In more recent times, there has been increased interest in literary statistics. Reference was made to quantitative statistical analyses in formulating psychological and stylistic laws, for example, on the length of phrases, on the distribution of phonetic accents, on absolute and comparative frequency of words, of parts of speech, or of phonemes (in the meaning of letters of the alphabet). There are scholars in the United States who have made important progress in this field.[2]

Therefore, the subject matter to be analysed is made up entirely of what can be transcribed from human speech into characteristic signs or symbols. It can be considered as having three levels. First of all, the word is the fundamental unit, and it is at the same time the graphic and semantic unit. Then there are sentences and phrases composed of more than one word. On the other hand, there are elements of each word, such as roots, prefixes and suffixes. In the same way we speak of atoms, molecules, and electrons.

Thus, it would be interesting to know which are the words used by a person or an epoch or a language. How many are there? To what radicals can they be reduced? What is their frequency? Their length? What are the rhythms of their accents? How are words distributed in phrases? What are the fundamental structures common to the phrases? There are many such questions.

The problem requires the searching, separating, arranging, correlating and study of a large number of small elements, tens of thousands of words, hundreds of thousands of letters. Such investigations must be repeated many times on the same material with different emphases and for diverse purposes.

For these studies we must record every unit of information on a free and manageable medium, such as a card. Punched cards permit multiple coding of the same information, and they can be sorted and re-sorted rapidly. In addition, the great—even enormous—quantity of cards to be handled, and the possibility of making automatic printouts directly from the cards, dictated the choice of machine-sorted punched cards. Among these I have finally chosen the IBM system, not only because Providence obtained for me the full collaboration of this company, but also because of the great flexibility of the system, because of the developments the company foresees in the near future because IBM has been developing machine methods for scientific documentation.

---

[2] For a list of U.S. scientists in this field, see for example, Guiraud, Pierre. *Bibliographic critique de la statistique linguistique*. Revisée et complétée par Thomas D. Houchin, Jaan Puhvel, et Calvert W. Watkins, sous la direction de Joshua Whatmough. Utrecht, Editions Spectrum, 1954; xix, 121 p. (Comité international permanent de linguistes. Publications du Comité de la statistique linguistique, 2).

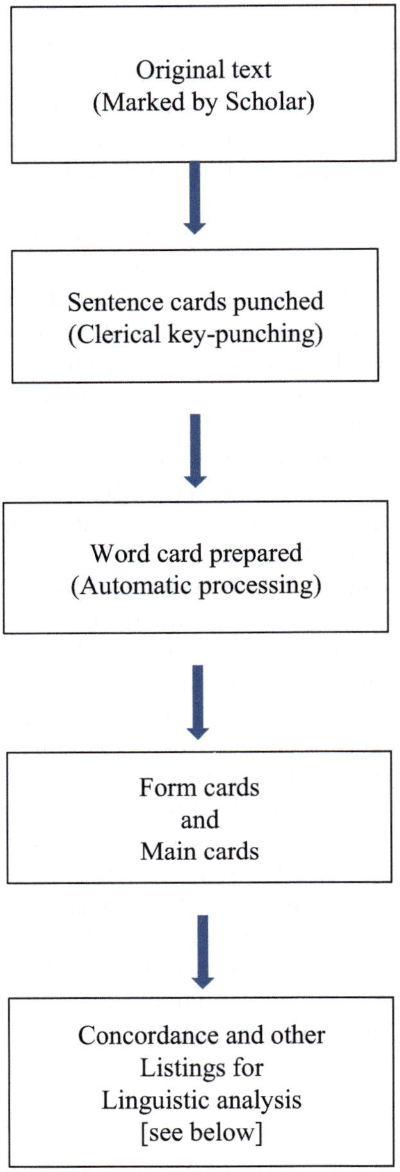

Fig. 3.2. Summary of operations.

I will now recount all that I have done, and all that there is still to do. I will use a flow chart (Figure [3.2]) as the basis for my discussion. This was first prepared at IBM in Milan by Mr C. Folpini and then completed in the offices of IBM in

New York with the assistance of Mr P. Tasman.[3] It will be evident that the process so described could be shortened considerably, if it is sufficient to obtain simpler results. However, it is useful to give the whole picture to show how much can be obtained if desirable.

Analysis of Words in a Text

I will concern myself here with the principal task, in terms of size and value, of linguistic investigation, which is making a concordance of a continuous text. It will not be difficult to apply these techniques to other problems, such as analysing the answers of questionnaires, or the words found as items in a glossary.

The scholar marks the text to indicate how it should be recorded on the cards, noting the beginning and end of paragraphs and of sentences with their appropriate references. Also, he distinguishes words quoted by the author from other writers, from the author's own words, etc. Where it is important not to mark directly on the text so as to deface it, a sheet of cellophane may be placed over the page and appropriate markings made on it with washable inks.

Each line from the text is punched into a card, one line after the other, each with [an] identifying reference to its place in the text. The maximum number of columns available for this punching is determined in advance, depending on the format wanted for the concordance. Thus, a maximum sentence length is established. Words are never split between cards; rather, a word is started on a new card if it will not fit on the preceding one. See Figure [3.3].

The problem of verifying the punching is an important one, because an undetected error will always be repeated. Errors can be detected in the usual way by checking the card on the verifier, or by proofreading cards on which the punching has been interpreted.

---

[3] "Literary Data Processing," P. Tasman, *IBM J. Research & Development*, 1, No. 3, 249–256 (1957); "Literature and Document Research Automation," P. Tasman, *Automation Systems*, Engineering Publishers Division of the AC Book Company, Inc., 1958; 61–72.

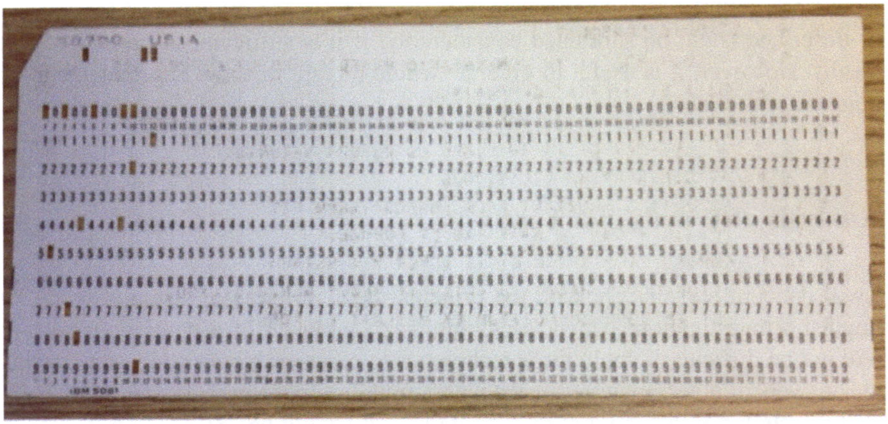

Fig. 3.3. Punched card from the *Index Thomisticus* project. Note: as the resolution of the image that appeared in the original article was so low we have instead inserted an image of another punched card made by the *Index Thomisticus* project (currently uncatalogued, Busa archive).

These operations produce the first or fundamental group of cards called *text cards*, or on the flow chart, the *sentence cards*. With this first and only data transcription, it is possible to accomplish mechanically, speedily, and accurately all of the most diverse and complex analyses.

To divide the sentences into single words, each on a separate card, the following methods could be used:

(a) punching each word from the text onto a separate card.
(b) simultaneous use of the sorter and reproducer equipment as described in the small volume, *Varia Specimina Concordantiarum*.[4]
(c) using the Cardatype, recently developed by IBM.

The use of the Cardatype offers the advantage of preparing typed copy of the text while punching the individual word cards. Thus, the context of the word is printed on the reverse side of each card.

This operation results in a second set of cards, the *word cards*. Each word is accompanied by reference to its place in the text. This file contains as many cards as there are words in the text. The word cards are alphabetized, using the sorter. Mechanical alphabetizing requires two passes of the cards through the machine for each column sorted. Thus sorting 100,000 cards containing words of 10 letters means, in effect, passing 2,000,000 cards through the machine. Depending on the model sorter used, from 30,000 to 60,000 cards per hour can be sorted. Therefore, it could take from 35 to 65 hours, approximately, to accomplish the alphabetization. In other words, the machine would alphabetize from 1,500 to 3,000 words of 10 letters in an hour.

[4] Roberto Busa, *Varia Specimina Concordantiarum*, Fratelli Bocca, Milano, 1951, 180 pp.

The operation can be shortened by several means, e.g., by separating first the shortest (one- and two-letter) words, then the next shortest words, and so forth. The shorter sets so divided are then inserted into the alphabetically sorted sets of longer words. The final result will be that all the words of the text are alphabetized and all identical words are grouped together.

Each group of identical words is given the same sequence number.

The accounting machine is used to print a list of all the words from the word cards. It is possible to prepare an abridged list on which only the different words appear. The machine will also print the total number of cards on which each different word appears. This gives the frequencies with which each different word appears in the text. The accounting machine may also be set to print the code number identifying the different forms of individual words.

When the summary punch is connected to the accounting machine, a third series of cards can be obtained while the list described above is being printed. These cards, called *form cards,* or *different word cards*, contain each different word with a number indicating its position in the alphabetic sequence and the total frequency of its appearance. Such a group of cards is not necessary for ordinary concordance work but it may open up the possibility of different and new investigations. Such cards, in fact, contain the summary of the author's vocabulary and can be analysed indefinitely. Mark sensing techniques are particularly useful for this (as will be discussed later).

In the list of words described above, the words are considered according to their graphic structure. Therefore, the scholars must separate cases of homographs, which turn out to be quite frequent; dismember words that comprise prefixes and suffixes each having a proper function also when isolated (such words may be considered as two words rather than one); assemble the separate words that are in reality just one verb form; and finally, regroup under the functional semantic unit all the diverse forms a word assumes according to case, tense, mode, etc. Such work requires the competent responsibility of the scholar and it cannot be accomplished by machine. However, once such classification has been made, mechanical recognition of different forms of the same word could follow.

The *main words* must be punched one per card with a special layout designed to accomplish the functions of these cards. They must also be arranged in alphabetical order and numbered progressively. Such a number-code may be added to the cards of the other two sets, word cards and form cards.

In the form that I have summarized there is obtained from one initial punching of the text, four groups of cards. They are: the text cards and the word cards, that contain all the words of the text and represent two new editions of the entire text; and the form cards and the main cards, which constitute two summary indexes of the vocabulary used in the text. The first lists the words grouped according to graphic form, the other lists the same words arranged according to graphic-semantic units.

Note that the word cards are accompanied by the elements necessary to characterize their individuality. The dissociation of the text into its first elements is, therefore, entirely reversible: it is always possible to reconstitute the text from these elements. Such proper determinations, reserved and exclusive of each single word, are its various codes. In fact, every word is coded as to its location with the reference and with the number of its position in the text; it is coded as a morphologic unit with the progressive number that it acquires in the first alphabetic sequence; it is coded as a semantic unit, with the progressive number that it has in the last alphabetical order.

Besides, it is accompanied by its context. Such context may be punched or printed. It may be punched on the same card on which the word is punched; a condition, however, that is restricted to about 50 to 60 letters and spaces, i.e., columns on the cards. It may be punched on another card, and then occupy even 70 or more columns, according to the length of the reference; such a card would then be the same text card. The context may be also printed in the spaces between the punched holes, and then it can be extended to twelve lines, and contain from 80 to 120 words; ample context that would almost always be sufficient to individualize the significance of the word without requiring the scholar to make frequent recourses to the printed text.

Finally, every single word can be accompanied by the first or last letter of the preceding word, by the first or last letter of the following word, besides the preceding quotation and the following quotation.

There also remains in the word card or in the form card or in the main card sufficient space for additional classifications to be applied, for example, manually by means of mark sensing. The researcher can make a symbol that tells what part of speech the word is, on what syllable the tonic accent goes, what is its length in letters or in syllables, and other things, too.

So, the resolution of the text in its first elements is completed. The four groups of cards represent the material suitable for whatever investigation in whatever direction: investigation that, in its quantitative aspect involving large numbers of small elements, is accelerated enormously by mechanization and rendered more accurate, more certain, and absolutely complete. The same cards will serve the most diverse analyses, because once the cards are selected according to a determined order, they may be brought back rapidly to their first order and subjected to new research.

One can obtain, from the one and initial punching of the text, various listings as summarized below.[5]

(1) The general catalog of vocabulary of the author, richer in prerogatives and more abundant of context than the same monumental concordance of *TLL*[6] prepared in Munich, Bavaria.

[5] [Editors' note] A figure that was included in the original version of this text, depicting sentence cards, has been omitted due to the low resolution of the image. The text has been amended accordingly to remove references to that figure.

[6] TLL – *Thesaurus linguae latinae*, editus auctoritate et consilio Academiarum quinque Germanicarum Berolinensis, Gottingensis, Lipsiensis, Monacensis, Vindobonensis. Lipsiae,

(2) The listing of the cards in various forms, using the accounting machine at a rate of 4500 to 9000 lines/hour. For example:
   (a) The text cards may give a reprinting of the entire text.
   (b) The word cards may give the alphabetic list of all of the different forms under which the words used are presented in this text, indicating their frequency.

This *laterculum formarum* may be obtained immediately after the words have been alphabetized and arranged. But if the code number of the main word is required, it would be necessary to wait until after the *rationarium verborum* has been prepared. (c) The *rationarium verborum* or *formarum* would be the diagram, systematized and with frequencies, of all the same words regrouped according to their meaning, or more exactly, according to the identity of their functional elements. Such a list is the basis of the author's vocabulary. (d) It would be very simple to list an abridged *conspectus lemmatum.* (e) The *index verborum* will be the index of all the words, or rather of all the word cards, with the reference and arranged according to the *rationarium.* (f) The Concordance will be the same list with the single words followed by the notation as well as the reference. The context may be of one line only, whatever is deemed to be sufficient; but it may also consist of three or more lines; in this case the word in question will always be found in the middle line.

For a simple Concordance, preceded by the *laterculum* and *rationarium formarum*, the required time will, for the first phase, equal the time of one or two typings of the entire text; for the following phases, it will be possible to fulfil in one year that which would take 30 to 40 years of work with the old method. This is the case for the printed Concordance. When, however, one needs to compose a catalogue in which the words follow a context of 12 lines on single word cards, 20- or 30-years' work can be completed in one year.

In respect to the cost of the work, this much was made clear, on the basis of Italian prices rather than those in the United States: We compared, on the one hand, a form, composed correctly and paginated, ready to be put into the rotary press, and on the other hand, a Concordance obtained from the IBM accounting machine on mats adapted for lithographing, and ready to be passed through the offset system or any other system of lithographic reproduction. We did not include the cost of materials, paper, or zinc. A line of the Concordance prepared and tabulated with the IBM system costs about half what it would cost to set up a line with a linotype or monotype. The computation was made on the supposition that all the work is done in the IBM offices at commercial prices. The difference in the cost

---

Teubner, 1900. Current volumes read: *Thesaurus linguae latinae*, editus iussu et auctoritate consilii ab academiis societatibusque diversarum nationum electi. "The great dictionary of the language, in Latin, indispensable in the university or large reference library. Plans to record, with representative quotations from each author, every word in the text of each Latin author down to the Antonines, with a selection of important passages from the works of all writers to the seventh century." Winchell, Constance M. *Guide to reference books*. 7th ed.

will be more appreciable, if one realizes that the cost of conventional printing does not include the cost of preparing the Concordance; while the cost of the IBM listing also comprises all of the work and materials of preparation, such as punching, sorting, and reproducing the cards. In other words, the new method, at half the price required for the preparation of the printing of a Concordance, gives not only the matrices for printing, but also the entire catalogue in a flexible form always ready for new studies.

LATERCULUM Verborum

Numerorum qui singulasubsequentur verga primus frequentiam, alter cui lemmati in Rationario adunetur indicabit

1 A	2	1
2 AB	1	1
3 ACCIPITE	1	2
4 AD	4	3
5 AEMULIS	1	4
6 AGITUR	1	6
7 AGNUM	2	5
8 AGNUS	1	5
9 AMBIGITUR	1	7
10 ANGELICUS	1	8
11 ANGELORUM	1	9
12 ANIMOSA	1	10
13 ANTIQUUM	1	11
14 ASSUMITUR	1	12
15 AUDE	1	13
16 AUXILIUM	1	14

A. Alphabetical listing of words as they appeared in the text. (Note that a serial number precedes the word as listed. First number after each word indicates frequency of appearance in text and the second number refers to the "Main word listing.")

CONSPECTUS LEMMATUM RATIONARII

1 A AB
2 ACCIPIO ACCIPIS ACCIPERE
3 AD
4 AEMULUS AEMULA AEMULUM

5 AGNUS AGNI
6 AGO AGIS AGERE
7 AMBIGO AMBIGIS AMBIGERE
8 ANGELICUS ANGELICA ANGELICUM
9 ANGELUS ANGELI
10 ANIMOSUS ANIMOSA ANIMOSUM
11 ANTIQUUS ANTIQUA ANTIQUUM
12 ASSUMO ASSUMIS ASSUMERE
13 AUDEO AUDES AUDERE
14 AUXILIUM AUXILII
15 AZYMUS AZYMA AZYMUM
16 BELLUM BELLI
17 BENEDICTIO BENEDICTIONIS
18 BIBO BIBIS BIBERE

B. Main word listing. (Note sequential number of these main words and citation of various forms.)

RATIONARIUM VERBORUM

Post singula lemmata proprio numerate numero, vocabula singulanumerus praecedei quo in Laterculo primo continebantur, numerus vero subsequetur frequentiae singularis ac tandem collectivae.

1 A AB

1 A 2

2 AB 1

3

2 ACCIPIO ACCIPIS ACCIPERE

3 ACCIPITE 1

1

3 AD

4 AD 4

4

4 AEMULUS AEMULA AEMULUM

5 AEMULIS 1

1

5 AGNUS AGNI

7 AGNUM	2

C. Word index combining all A-words with corresponding B-words indicating—to right—frequency of occurrence. (Numbers to left refer to listing B.)

INDEX VERBORUM

1 A AB			
A	L	15	43
A	V	2	5
AB	P	6	35
2 ACCIPIO ACCIPIS ACCIPERE			
ACCIPITE	S	4	15
3 AD			
AD	P	4	23
AD	S	7	28
AD	V	1	3
AD	V	1	4
4 AEMULUS AEMULA AEMULUM			
AEMULIS	V	2	6
5 AGNUS AGNI			
AGNUM	S	2	6
AGNUM	S	3	9
AGNUS	S	21	66
6 AGO AGIS AGERE			
AGITUR	L	6	16

D. Word index combining word entries from A and B with citation of all occurrences in the text. (Numbers to left refer to listing B.)

CONCORDANTIAE

1 A AB

A SUMENTE NON CONCISUS				
	L	15	43	A
PROCEDENTI AB UTROQUE				
	P	6	35	AB
IN MORTEM A DISCIPULO				
	V	2	5	A

2 ACCIPIO ACCIPIS ACCIPERE
DICENS ACCIPITE QUOD TRADO
VASCULUM S 4 15 ACCIPITE
3 AD
AD FIRMANDUM COR SINCERUM
P 4 23 AD
AD LUCEM QUAM INHABITAS
S 7 28 AD
AD OPUS SUUM EXIENS
V 1 3 AD
VENIT AD VITAE VESPERAM
V 1 4 AD

E. Concordance. (Listing in main words in textual context with identification of location in text).

Research on the Structural Elements of Words

A statistical list of prefixes is already contained in the first *laterculum verborum*. It is also very easy to obtain a list of only the first three or four or more letters of words, with totals of frequencies of the single different beginnings of words. It is evident, however, that in this case the machine will list also the short words composed of less than five letters, unless the scholar prevents this by appropriate instructions to the machine.

It is possible to sort from the sets of word cards and form cards those words with particular combinations of initial four of five letters. One can also use the collator, in which a pilot card punched with only those letters that constitute the prefixes desired, instructs the machine to extract those cards that contain that composition of initial letters.

If cards are placed in the accounting machine to obtain the list of word cards or form cards, a summary punch can be coupled to obtain a series of cards that represent the various beginnings of words, accompanied by a code number (a serial number representing the alphabetic order) and by the total of the frequencies.

In like manner one can turn to the analysis of the endings. For this we need to punch the words so that the last letter of each word is in the same column. This can be done with the sorter, by separating the words by length, then reproducing all of the cards so that the last letter of each word will be found in the same column. This task is simplified by working from form cards. The words so punched are now alphabetized backwards. This is done by sorting first the initial letter of the longest word and then proceeding from left to right. (This in the reverse of the usual IBM alphabetizing procedure.) We now have the reverse index, in effect a rhyming dictionary. In this way we can list different endings of words indicating the number of frequencies. Also, we can pair the accounting machine and the summary punch to obtain a series of cards, which contain the endings of the words, in order to work only on these.

The calculation of the number of letters of a text or of a vocabulary becomes a very simple operation when those words are already punched in cards. One can, for example, use the text cards and explore every single column to sort the letters present in that column. Also, we can count each package with the card counter of the sorter, and write the sum totals. To the total of the first column we add the sum of the letters present in the second column and so on. This operation is facilitated by using a sorter provided with a counter, and even more by using the 101 statistical machine.

More work is necessary, but it is still extremely fast compared to manual labour, when one undertakes to analyse the distribution of letters, diphthongs, for example, in words. Such an inquiry, in fact, coincides with the search for roots of words. Such inquiries can be done by means of successive sortings. This could be shorter if the collator were used, with a pilot card. It would be shorter yet if the 101 statistical machine were used: this machine searches at the same time four different groups of three letters in each word of twelve letters punched in a card. The cards pass through the machine at a rate of 46,000 per hour. The machine does not select the roots only, but rather any combinations of letters specified.

The machine will also facilitate the preparation of materials for study of the distribution of tonic accents, of the proportion of use of the parts of speech and other things. For example: what percent of nouns, of verbs, of adjectives … or what is the predominant structure of the phrase: subject-verb-complement (or predicate), or instead, predicate (or complement)-verb-subject.

Mechanical Search of Phrases

The actual system as described permits searching for a particular phrase, if as emphasized above, the first letter of the preceding word and that of the following word were punched on the word cards. There is, for example, the saying, "*sotto questo punto di vista.*" Among all the cards that carry the word *questo*, the sorter separates those in which *questo* is preceded by *s* and followed by *p*. Among the words *di*, those are selected that are preceded by *p*, preceded in turn by *q* and *s*, and followed by *v*. These cards give the references to all the passages in which are found said sentence, even if the sentence is distributed on two successive cards.

It must be noted here that all that has been said is not necessarily limited to Latin characters. The machines can be provided with any alphabet, and also for Arabic and Hebrew which proceed from right to left. Any series of symbols, signs, or ciphers may be applied to the machines.

The analysis requirements for most texts necessitates the use of punctuation and diacritical marks. The IBM accounting machines such as the 402, 403 and 421 can be utilized for these marks by substituting for the Arabic numerals the desired symbols. This is also the case where card interpretation is required on the IBM 552. Card punches may be modified by substitution of suitable key tops.

For those symbols that should accompany the word, the 12 and 0 zone punch positions of the card should be reserved for use with all IBM accounting machines except the 407. For example, the apostrophe for the articles with elisions in Italian and for the genitive in English, and the point (or period) for the abbreviated words must accompany the word even when it is isolated in the word card. Thus, the German *das ist* becomes abbreviated d. i., but is punched like d-i-; and is also listed as d-i-.

Adoption of the diacritical symbols and of interpretations offers major possibilities through the use of the IBM 26 card punch and 407 accounting machine, where in addition to the numbers and the letters there are spaces for 11 special characters. Ordinarily such characters are accounting characters, but they can with moderate cost be substituted by signs used in linguistic or philologic studies.

Problems involving accents are more difficult to resolve by ordinary means. Some languages actually present a considerable problem: one thinks of modern French and classic Greek. When special characters for symbols of punctuation, diacritical marks and accents are required at the same time, the top space of a column is not sufficient, except for certain kinds of work and for some languages, for example English, Italian, or Latin. It is necessary then to consider IBM machines which use codes of punches consisting of combinations of two or more columns.

The Near Future

An application of punched cards that should be explored more fully is the automatic tracing of the variants of the same text. The first step in any critical analysis consists of comparing the results of the same analysis applied to a representative selection of various manuscripts or editions. The first line is written down, then the variants as encountered in the other copies. From this material the researcher determines whether the first word is an authentic word of the author. It would be possible to devise a process as follows: Punch line after line for all the editions judged to be representative, one line per card with the reference and proper symbol for each edition. The sorter will assemble all the first lines according to the symbols for the editions, then all the second lines, then the third lines, and so on. The cards ordered in this way are fed to the accounting machine, set to print the group of first lines, then to leave a space before printing the group of second lines, etc. The machine can be so set that for every group, it prints the first line in its entirety, and only those parts of the following lines that are different from each preceding line. Probably it will also be possible for the machine to print, for all of the successive versions of the same line, only that which is different from the first line. Among the graphs is a model showing the tabulation for punching of a set of hypothetical variants of a verse from Dante (see below).

1 POCA FAVILLA GRAN FIAMMA SECONDA
2 POCA FAVILLA GRAN FIAMMA SECONDA
3 POCA FAVELLA GRAN FIAMMA SECONDA
4 POCA FAVILLA GRAN FIAMMA SECONDA
5 POCA FAVILLA GRAN FIAMMA ASSECONDA
6 POCA FAVILLA GRAN FIAMMA SECONDA
7 PRIMA FAVILLA GRAN FIAMMA SECONDA
8 POCA FAVILLA GRAN FIAMMA SECONDA
9 POCA FAVELLA GRAN FIAMMA SECONDA

Fig. 3.4. Tabulations of a Set of Hypothetical Variants of a Verse from Dante (*Paradiso* I, 34)

Such a process presents problems of cost. It must be established whether the cost of punching and verifying the same text as many times as there are variations is compensated by the speed and certainty of the subsequent analysis and by the fact that the operation produces text cards for linguistic analysis and the preparation of a concordance of the text.

There are technical problems yet unsolved. The principal problem has to do with prose writings. Due to the fact that the machine exercises control on each column of the card, should a single letter be left out of one line as punched on one card, then the remainder of the text on other cards would be displaced by one column and would as a result be printed as a variant. Such difficulty does not exist for writings in verse, each line of which is started on a new card. For prose works the problem might be resolved by using punched tape or the electronic computer with sufficient memory capacity.

Another development is the printing of the differential context on the back of the word cards. If on all the cards for the words in a given section of text there is recorded the same context, then the first word does not have any preceding context and the last word does not have any following context. As the system is developed there is a need for the context to be printed in such a way that the word punched on the card will be found in the centre-line of those printed. Thus, for all of the words of line 20, the text would begin on line 14 and end on line 25; for the words of line 21, on line 15 to line 26; for the words of line 22, from line 16 to line 27, and so on. This problem is not exactly one of machines for punching cards, but rather one of duplicating machines. However, it is desirable that the processes be linked with the punching on the same cards.

In addition, two parts of the process need to be accelerated: reproduction from the text cards to the word cards, and alphabetic sorting. These two phases notably affect the time and cost of the analysis procedure. Difficult problems of these types obviously do not occur in the ordinary industrial and statistical applications of the IBM system. It is also obvious that, for this reason, an answer to such a demand will come in the future, when research work on linguistics has justified the cost of constructing new models of machines or at least adapting models already in use.

The possibility of punching text cards automatically, starting with the examination of the text by means of a photocell or by other means exists but as yet there are no practical methods for carrying it out. Naturally it will be gratifying when such techniques are operational.

For preparation of concordances, those most valuable kinds of philological studies, perfection of the method will be achieved when it is possible to have cards that carry three lines of context punched on the same card. The amount of context carried in a line of 80 columns is not sufficient in a printed concordance prepared automatically. For the most part it is necessary to have a context of three lines, so that the word in question is always in the middle line. As already described, it is possible to obtain such printing

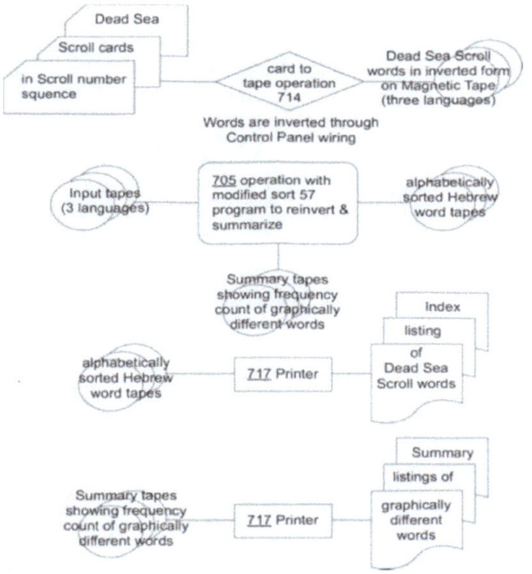

Fig. 3.5. Simplified block diagram of "Dead Sea Scrolls" processing on EDPM equipment

even with standard machines, but repetitive use of the sorter and collator is necessary as well as extensive operations with the accounting machine. When such extensive operations can be avoided by using cards containing three lines of context, or some equivalent means, the mechanization of linguistic analysis can be said to have reached that stage where substantial change will not be required for some time.

When punched card systems operate so that the cards are passed through the machines along the cards' long axis, which is perpendicular to the axis of motion through present day machines with exception of the punchers and verifiers, then it would be possible to work on the basis of successive circuits, like a telephone centre. The demands of linguistic analysis would then be satisfied even more completely, faster, and more economically.

Such observations on desired developments should not overlook the fact that even with standard machines the punched card system permits more extensive, more certain, more advanced and more economical studies than would have been possible except with the patient work of many men.

For similar work in the preparation of concordances by machine, the reader is referred to the work of Reverend James Ellison in preparing a concordance of the Bible by means of a computer. See, for example,

"According to Mark 4.-," *Time*, vol. 64, August 9, 1954, pp. 68–9
Soule, G. "Machine That Indexed the Bible," *Popular Science*, vol. 169, November 1956, pp. 173–5
"Bible Labor of Years is Done in 400 Hours," *Life*, vol. 42, February 18, 1957, p. 92
Cook, C. M. "Automation Comes to the Bible," *Christian Century*, vol. 74, July 24, 1957, pp. 892–4

Fig. 3.6. Father Roberto Busa comparing the words of a modern scribe—the printing unit of an IBM 705 computer—with those written two thousand years ago by scribes of an ancient Hebrew sect living near the Dead Sea (27/09/56, New York; Busa Archive #0032)

Appendix

Work is now in process in Gallarate and in New York in applying the method of literary analysis described here to the task of cataloguing the "Dead Sea Scrolls." The nearly thirty thousand words under study were punched into IBM cards. A card was punched for each word, indicating its exact location and distinguishing characteristics. The entire set of cards was converted to two reels of magnetic tape by the IBM 705 computer in approximately two hours. See Figures [3.5] and [3.6].

In order for the IBM 705 to perform the indexing of the "Dead Sea Scrolls," the following items had to be taken into consideration.

1. Card input requirements
2. Printed output requirements

Since Hebrew words are right-most justified, read and interpreted, special considerations had to be dealt with prior to obtaining the desired results. For the input, the Hebrew word cards are initially converted to tape in such a fashion that the words will be recorded on the magnetic tape in an inverted form (left-most justified).

The Hebrew words range from 1 to 12 character positions.

The magnetic tapes, once created, are then loaded on the IBM 705 and with the aid of modified Sort 57 Program Deck, the following has been accomplished.

1. This program first sorts all these Hebrew words alphabetically and at the same time re-inverts them into their original form prior to writing them on the output tape.
2. An extension to the program provides for creating a summary word tape on which are written only those words which are graphically different from each other. This summary record will also show an identification serial number with the frequency of usage of each word.

Later, on an off-line basis, these tapes will be listed on a tape-to-printer operation. The total off-line printing time is five hours.

Chapter 4 The Main Problems of the Automation of Written Language

Roberto Busa S.J.

First published as: Busa, R. 1958. I principali problemi dell'automazione del linguaggio scritto. In *Atti della VI Sessione delle Giornate della Scienza – Convegno Int. sui Problemi dell'Automatismo, Milano, 8–13 Aprile 1956*, vol. I. pp. 831–841. Roma: Consiglio Nazionale delle Ricerche.

Editors' note

This article shows Busa's purview widen beyond the philosophical and linguistic analysis of Aquinas and related authors as he builds new arguments for the significance of his research to newly-emerging research areas like machine translation. He begins with an overview of approaches to the automation of written language that are being pursued by three sectors: library science, information science and machine translation. He presents these three sectors as all struggling with the same three problems: the mechanization of the alphabet; the codification of the lexicon and the expression of linguistic structures as mathematical formulae. Busa writes that the research that he has been undertaking (and which he describes as the most arduous of all) on the mechanization of linguistic analysis offers solutions to these problems in the form of the 'analytical, comparative and statistical' data that he has derived from the languages he is working on. This article includes early mentions of some of the research areas and questions that would especially occupy Busa in the years that followed. We learn of his burgeoning interest in machine translation. We also read of his growing interest in exposing what the machine could not do and in the valorization of this perspective as deeply humanistic. As we shall see, the idea that: "It is the despised machine that repeats to us the invitation 'know thyself still more profoundly, scientifically and humanistically: study your speech'" is one that he would continue to return to even in his final publications.

Abstract

1. The automation of written language (every automatic and useful transformation and elaboration that is not merely reproduction or transmission) is required nowadays by the expansion of the needs and size of that public utility represented by the communication of scientific information. The three sectors that deal with it are: *a*) *library science,* with research into the automation of the catalogue; *b*) *information science*, with research into the automation of the bibliography; *c*) *mechanical translation* or automation of translation.
2. For each of these sectors we are going to list: *a*) the main *achievements*; *b*) the main review articles and *publications*; *c*) the main research centres.

J. Nyhan and M. Passarotti (eds.), *One Origin of Digital Humanities*,
https://doi.org/10.1007/978-3-030-18313-4_3

3. The main problems common to these three sectors concern: *a*) the mechanization of the alphabet; *b*) the codification of the lexicon; *c*) the expression of linguistic structures in mathematical formulae.
4. The present state of research in these directions is described, and the related research centres are listed.
5. The mechanization of linguistic analysis, with which the present author is concerned, prepares automatic methods of rapid, multiple and extensive surveys of linguistic material that are analytical, comparative and statistical. The techniques that it develops aim at the increase of speculative studies in language (linguistics, psychology, philology and philosophy) and consequently at contributing to the solution of the aforementioned problems.

1 Introduction

I speak of *language* in the restricted sense of words properly spoken, *id est,* of the sequences of graphic symbols that signify awareness of objects. Therefore, I do not speak of those symbols that are numbers, nor of other symbols, such as, for example, the data on hospital files and those of H.M. Nautical Book, which are merely calculated and printed by using punched cards.

I speak of *automation* not in the sole sense of reproduction or transmission of information or messages. Therefore, I do not give a summary of the whole range of these techniques, which extend from punched tapes of the monotype and the teleprinter, to the *Lumitype* of the Graphic Arts Research at Boston USA, the *Ultrafax* in experimental use in the Library of Congress at Washington D.C.; and from the microfilm to the microcard and the most recent minicard of Kodak. I mean automation as processing of linguistic data, processing that can be reduced to two different denominations of selection and comparison. "Today, the sectors where language, in the sense said above, undergoes such automation are four:

1. librarianship for the automation of the catalogues
2. documentation for the automation of bibliography
3. machine translation
4. linguistic analysis.

The first three sectors overlap or at least intersect in various areas of their function.

2 Automation in Library Organization

I do not deal here with data processing for library administration, such as the recording of loans on punched cards. However, I do outline the enormous requirements of the catalogues of great libraries. There are problems of updating: putting new titles in alphabetical order in different catalogues; integrating the catalogues of different libraries into "unique catalogues" that are regional, if not national. And then there are problems of use: how to make the most complete and economical selection of material for a particular inquiry; how to reproduce lists, whether partial or complete, of the bibliography cards so chosen.

The possibility of future mechanization depends especially on *cost*, for it would cost an enormous sum to put all the existing card-indexes of a great library onto punched cards; not to mention the fact the catalogue should always maintain its essential characteristics, *viz.* ease of access for cursory and detailed consultation.

As far as I know, there are two places where this is being carried out:

1. The Library of Congress is compiling the *catalogue of titles* of newly published periodicals on IBM punched cards and with the IBM 407 tabulating-machine is printing the periodical bulletin *"New Serial Titles";* thus, the mechanization is restricted to an extremely specialized catalogue of the library and, at least at present, it does not extend to lists of previously published periodicals.
2. A bolder undertaking owing to its large scale is represented by the *Unique Catalogue of Italian national libraries* mechanized in Rome with Powers cards under the direction of Prof. Aldo Ferrabino.

An unsolved problem in these undertakings is the limited number of available symbols (36–38 in the Powers system, 47 in the IBM American model); this imposes considerable sacrifices of punctuation, diacritical marks, capital letters, and letters of other languages.

3 Automation in Information Science

Library catalogues contain only titles and mainly those of volumes. However, information science, which was previously a branch of library science, has nowadays developed into a separate discipline. Information science seeks to make the contents of a publication, as well as its title, selectable, and deals with articles in periodicals and patents as well as volumes.

The need for automation in this area caused a development known to all, that of the index. Existing indexes, by their very nature, always give answers to requests that are already known, and these indexes can be very complex and detailed. But how does one look for an answer to a new interest? In other words, what type of index meets the needs of research as strictly defined?

Automation has been established on the following principles:

1. that only one copy of a document is kept;
2. that a summary, the *abstract* of it, is made;

3. that this summary is consequently codified with graphic symbols that correspond to the classification of the objects;
4. that the order in which documents processed in this way is kept is not absolutely necessary for research, but merely useful for rapid consultation.

Symbolic codes are operated by a machine in different ways and with different techniques according to their nature: they can be *holes* in paper, or *magnetic impressions* on a tape varnished with iron oxide. A very rapid selection of the documents kept will be completed by comparing all of them with a *pilot card* or something that corresponds to one in which the indicative elements that are being researched are matched in sequence.

To the best of my knowledge, the already accomplished automations are nearly all in the field of chemistry. Abroad there is the Centre for documentation of the Gmelin-Institut at Clausthal-Zellerfeld in Germany, which is organized and directed by its president, Prof. Erich PIETSCH; its systems employ punched cards with perforations in the margins, and so can be used manually (Keysort) and by the IBM electro-countable cards. From here the method passed, always in the sense of information science as properly defined, to various German universities and academic institutions and to various firms such as Basf and Bayer.

In the United States the work of Prof. James W. Perry is remarkable and ground-breaking. He first taught at the MIT of Cambridge, Mass., then he did research at the Batelle Memorial Institute in Columbus, Ohio, and is now Director of the Center for Documentation and Communication Research in the School of Library Science of the Western Reserve University in Cleveland, Ohio. He used IBM machines right from the start, and then special models of them, in which exploration was calculated with a photoelectric cell and the cards were explored along the longest part, i.e. one column after another, as if they were pieces of tape; at the moment, he is having his own machines constructed. Today such methods are being applied to the documentation of metallurgy.

Still in the United States, the day before yesterday, 11th April, at the annual Congress of the American Chemical Society in Dallas, Texas, some “mechanical patent searching equipment” was announced for the automation of research into the information contained in the patents deposited in the USA by the authors B. E. Lanham, J. Leibowitz, and H. R. Koller in the United States Patent Office. Here I omit other examples, such as the work on mechanization of medical bibliography by the young Eugene Garfield, which I refer to below

3.1 Bibliographical Outline of Documentary Automation

The picture, which I believe to be complete, of the situation of automation in the field of documentation is given by two books:

R. S. Casey & J. W. Perry – *Punched Cards. Their applications to science and industry* – New York City, 1951. (This has had three reprints. The authors ask me to ask you to collaborate on the second edition, which should come out in a few months' time).[1]

J. W. Perry, A. Kent & M. M. Berry – *Machine Literature Searching* – Interscience Publ. (USA) 1956.

The following articles can also be profitably consulted:

E. Pietsch – *Dokumentation und mechanisches Gedächtnis zur Frage der Oekonomie der geistigen Arbeit* – Arbetitsgemeinschaft für Forschung des Landes Nord-Rhein-Westfalen, Heft 38.

E. Garfied & R. Hayne – *A National Science Intelligence and Documentation Center* – Report given to the Symposium on Storage and Retrieval of Scientific Information at the annual Congress of the Amer. Ass. for the Advancement of Science, 28th December in Atlanta, Georgia.

E. Garfield – 'The preparation of printed indexes by automatic punched-card techniques' – in: *American Documentation* VI (1955) 2, pp. 68 ss.

V. P. Cherenin – *Nokotorye Problemy Dokumentasil i Mekhanizatsilia Informatsionsnnykh Poiskov* – Inst. Of Scientific Information of the Academy of Sciences of the USSR, Moscow, 1955; (English version by R. G. LIEPINA in the Library of Congress in Washington, D.C.)

4 The Automation of Translation

Two factors affect the need for mechanical translation. The first is the *cost* of the translation performed by the human translator: it has been calculated in the United States that a good quality translation costs about 12–14 Italian lire per word; even an inferior translation will never cost less than 4–5 lire per word. The second is that the greater part of tests so far carried out or attempted have been on the automatic translation from Russian to English, attesting in such a way the underlying need for *national defence.*

Studies in this field have been carried out by the following experts:

- In England: A. D. Booth of the Birkbeck College Computational Laboratory of the University of London.
- In the USA: W. N. Locke, V. H. Yngve of MIT in Cambridge, Mass; and Leon Dostert of the Institute of Languages and Linguistics of the School, of Foreign Service of the University of Georgetown in Washington D.C., which is run by the Jesuits.

Recently I have been given information about an experiment in mechanical translation carried out by the Soviet Academy of Sciences, but as yet I have not received any documentation. In Italy Dr. S. Ceccato deals with this in his journal "*Methodos: Cibernetica e linguaggio*", as do other scholars here present. I know

[1] Here Busa is presumably addressing the delegates of the conference to whom he originally delivered the speech that formed the basis of this paper.

of only one successful experiment that has been made public up to now, that carried out by Prof. L. Dostert at New York in January 1954 with the electronic calculator of the IBM model 701, accounts of which you should certainly have seen in the press.

The general situation in mechanical translation is fortunately easily accessible to-day thanks to the book: *Machine Translation of Languages*, edited by W. N. Locke and A. D. Booth in 1955 (pub. by J. Wiley, New York City; Chapman and Hall, London); the bulletin "*Mechanical Translation*" published by Prof. W. N. Locke and others of MIT, Cambridge, Mass.

The techniques of mechanical translation can approximately be described as revolving around the following formula: *a machine in which there exists, either actually or potentially, connections between the graphic-semantic sign of each word in a language and the graphic-semantic sign of the corresponding word in another language.* When I say actually or potentially, I mean at least the possibility of combining automatically the elementary and irreducible morphemes of the source language that are actually present and connected with the corresponding, elementary and irreducible morphemes of the target language. I say the graphic-semantic sign because it is not a matter of translating just the sign but primarily the meaning. The machine must produce in the target language either the single word that means what is meant by the word in its context in the source language; or at least, in the case of homography, all the different words which in the target language can mean what is meant in different contexts by a single word of the source language. Such techniques require, on the one hand, that the machine is equipped with a memory and mechanisms of comparison, and on the other, that the language has already been expressed in mathematical formulae that can be mechanically operated.

The main difficulties come respectively from the limitations of a machine's memory and the size and complexity of linguistic facts, together with the still rudimentary knowledge that we have of them: it is enough to think of the problems raised by homography, by the different inflections a word can have, by syntax, and by idioms and metaphorical expressions. For this reason, the techniques are developing in two directions: one aims at making the machine more powerful, especially in its memory, the other at simplifying the language, either the source language (e.g. with *Basic English*) or the target language.

Thus, next to the ideal of a mechanical translation properly defined, which also comprises the translation of the syntax of the period, the function of the *mechanical dictionary* becomes more modest and yet more practical. It would give the literal translation of one word after another (giving all the possible correspondences of the homographs) and ignore the syntax. Experience would show such a rough version to be sufficient to understand the general sense of the text to determine whether it deserves a more complete translation by a human translator.

5 Automation of Language Analysis

The need for such automation comes from the need to develop analysis of language. Such analysis is most exhausting because of the mass and fluidity of the material and because of the infinite varieties of research to be carried out upon it.

If pure sciences can be distinguished from applied sciences, we should be able to make a list of those pure sciences that today need a greater abundance of, and easy access to, the documentation of the material that they investigate. They are: *linguistics* (*glottologia* in Italian) and *semantics*; *philology* and *grammar*; rational, experimental and medical *psychology*; and *philosophy*, because the characteristic method of metaphysics as interpreted by Christianity is a reflection on how much is common certainty to all human beings, whether professional philosophers or not. Certainly, common to all is what we might define as power or critical need, or else the possibility of understanding or of judging. Reflection upon it happens just by studying those means of expression that represent communication. Finally, even *theology* as properly defined, i.e. the study of the mystery of God, precisely because the mystery is revealed to us—and it could not be otherwise—with words that all can understand, i.e. with words that signify concepts which everybody has, and their juxtaposition tells us where the mystery is.

Among the applied sciences, which need the development of linguistic analysis, it will be sufficient to recall that mechanical translation needs precisely a larger knowledge of the recurrences of the language: what are the *homographs* of a language? What lexical context is characteristic of a given meaning of a *homograph*, so that the machine can recognize the exact meaning from the context? What words end in *-mente* and what in *-one* etc.? Such a need is met by the establishment of *mathematical formulae*, which can express the recurrences of linguistic structures in numbers or at any rate in symbols.

Therefore, two phases can be outlined:

1. the mechanization of the operations that can be carried out through the use of formulae already acquired;
2. the automation of the formulation of such formulae in the first place. In fact, it is obvious that these will become more effective in accordance with the extent, depth and variety of the research carried out to get to the point of inducing the general formula.

Corresponding to the first function, on the one hand, is *symbolic logic*, *operational mathematics* and *propositional algebra*, while on the other, is literary statistics or stylometry. In Italy this latter is dealt with by Prof. Marcello Boldrini in his *Manuale di statistica.* A valuable survey of the state of these very highly specialized studies is provided by the *Bibliographie critique de la statistique linguistique*, published by P. Guiraud (Utrecht/Antwerp, 1954) with the assistance of specialists from Harvard University, USA. For an essay see Victor Garcia Hoz—*Vocabulario usual, común y fundamental*—Madrid, 1953. *Cryptography* too might be mentioned, but is it not easy to give a summary of it, because by its very nature this discipline seeks to be as secret as possible.

In the area of *operational mathematics* automation appears as a “logical machine” that resolves logical functions with the process of tables of truth. Universal numerical calculators could already fulfil this function, but these were constructed out of mechanical parts; among the most recent there was the machine of Burkhardt and Khalin. In the past there were those of W. S. Jevons in the last century and earlier still, the *ars combinatoria* of the polygraph Ramon Lull, a monk who lived at the end of the XIIIth century.

My contribution, however, deals with the second of the two functions mentioned above: the development of *operational methods* that permit research into the first numerical proportions intrinsic to language. By taking advantage of the general card index of all the words in the works of St Thomas Aquinas and using the IBM systems, I am engaged in working out techniques that allow one, rapidly and on a large scale, to isolate, calculate, and codify the presence and proportions of frequency of words (distinguishing and separating inflections, homographs, compound words,...), morphemes (roots, prefixes, suffixes,...), syllables, letters and phonemes, accents, distribution of the parts of speech, length of sentences and phrases, etc..

Up to now I have examined texts of a maximum of 50,000 words. I have now completed the punching of the 220,000 cards that represent all the lines of the *Summa Theologiae* of St Thomas. My job in the next few months will be to split them up automatically into the 1,600,000 words that they contain: words to be arranged alphabetically in two directions, from the right and from the left, in order then to submit them to more varied statistical investigations. A slightly more distant programme is to carry out the work on something that contains four or five million words. Such an experiment corresponds to the well-known transition from laboratory tests to mass production. In fact, it happens that I come up against the problems posed by the simultaneous manoeuvring of blocks of hundreds of thousands, if not millions, of cards. This explains simply why in the near future I intend to occupy myself with the new electronic calculators with magnetic tapes (the Electronic Data Processing Machines, EDPM, of the IBM of the 600/700 group) in order to examine precisely how far such machines represent a saving, given that they eliminate the mechanical passage of the cards through the machines. Indeed, in these machines the cards are replaced by a magnetic tape, the unwinding of which is the only mechanical operation required, while all the other operations are carried out electrically.

Others, too, are working in this field: Dr W. Ellison of Winchester, Mass., USA, has computed the variations of the Greek codices of the New Testament with Mark IV of Harvard and at the moment is preparing his concordance of the English Bible, for pastoral, rather than scientific, purposes, with the UNIVAC of Remington Rand.

6 Conclusions

In conclusion, the problems posed to-day by the needs of the automation of written language can be summarized under three headings:

1. the machine alphabet;
2. the codification of the lexicon;
3. the mathematization of language.

6.1 The Machine Alphabet

This basically means adopting a new form of writing which is operable on a machine: something that more resembles cuneiform impressions on clay than the line of ink on paper! That is, it is necessary to find symbols of thought that can be determined by mechanical functions, which can thus be a characterized impulse or a condition characteristic of a characterized impulse. Examples include the holes on a tape or a card, the graphite marks on the "mark sensing" of the IBM, the black and white pictures on film and the magnetization of a tape varnished with iron oxide and other substances. The possibilities afforded by electrical and electronic techniques present two problems:

a) How to get from the usual sign to the mechanical one. Up to now the substitution has been performed by hand, for example, in punching machines, but techniques are being investigated to *transform automatically human graphic and phonic symbols into symbols capable of being operated by machines*. Probably, research into how to transform mechanically human written symbols into machine symbols is nearer to completion than research that is being industriously carried out into how to automatically transform human sounds into symbols operable by a machine.
b) Another problem is the adaptation of written symbols to phonemes as properly defined, in order to eliminate the differences between writing and pronunciation; one wonders whether it will ever be solved.

6.2 The Codification of the Lexicon

As I mentioned before, this involves adapting the distribution of the written symbols to classifications of the objects in question, with a view to simplifying the selection of what is involved in a particular subject. To give a made-up example: if all mammals are called by a word that has the syllable *ma* as its first syllable; if all dogs have the syllable *ca* as the second syllable; if the different breeds of dogs are distinguished by different syllables always in third place: the shepherd dog will be *pa*, the greyhound will be *le*, and the boxer *bo*: thus, we shall have *macapa, macale, macabo*, etc. When searching, if I am interested in everything about mammals, it will be sufficient to separate the documents according to the presence or absence of the binomial *ma* in the first syllable. If, however, I am interested in everything about dogs in general, I must select all the words that have *maca* as their first two syllables, and so on. This perhaps also means that to-day we have come to realize that we could have constructed our words more intelligently and less artistically, or less capriciously, which amounts to the same thing.

6.3 Mathematization of Language

With this phrase, I wish to express the concept mentioned above, namely that the structure of linguistic recurrences (language consists of multiple entities that are repeated and therefore countable...) in their widest sense (grammatical, syntactical, stylistic, graphic and graphic-semantic) can be expressed with mathematical formulae. How many thinkers have jumped for joy at discovering the relationship between numbers, shape, and beauty!

Such a discovery of formulae that are in continuous, active operation is required by the needs of the automation of linguistic analysis, an automation which is in turn required by the development of culture.

This leads me to conclude that the automation of written language awaits some technical development, but it also expects much more from the spiritual industriousness of mankind. The machine warns us that we are not humanistic enough and, although we speak, we are not able to explain how we speak. It is the despised machine that repeats to us the invitation "know thyself still more profoundly, scientifically and humanistically: study your speech". The automation of written language thus promises an increase in spiritual education.

Chapter 5 The Work of the "Centro per l'Automazione dell'Analisi Letteraria" in Gallarate, Italy

Roberto Busa S.J.

First published as: Busa, R. 1961. Les travaux du Centro per l'Automazione dell'Analisi Letteraria de Gallarate, *Cahiers de Lexicologie* 3, 64–68.

Fig. 5.1. Busa, on 08/10/61, shown with files of punched cards, in the midst of moving the operations of CAAL to its new location on via Galileo Ferraris, 2, Gallarate (Busa archive #0428).

Editors' note
In this text, Busa presents an overview of the work, organization and aims of the *Centro per l'Automazione dell'Analisi Letteraria* (Centre for the Automation of Literary Analysis; CAAL), which he set up in Gallarate, Varese in the mid-1950s.[1] In the early

[1] It is difficult to identify when exactly CAAL was set up, as Busa gives different dates for its founding in different articles, sometimes stating that it was founded as early as 1949 (Busa 1965 p.64). The more plausible date for its founding is c.1956. The earliest photographic evidence of

J. Nyhan and M. Passarotti (eds.), *One Origin of Digital Humanities*,
https://doi.org/10.1007/978-3-030-18313-4_4

years of the centre its operations were carried out in various premises in Gallarate, which were loaned to Busa by local Gallarate businessmen who were associated with the "Finance and Administration Committee" of the *Index Thomisticus* (Busa 1980, 85). Towards the end of 1961, the year that the article below was published, CAAL would move to a new location at via Galileo Ferraris, 2, Gallarate. This former textile factory had the space to house the staff who occupied the posts that Busa lists below, namely the philological staff, the secretariat and the laboratory workers (see Jones 2016, 120–123)

When Busa wrote the article below, the centre consisted of "30 units". The word unit should presumably be read to mean individual. Busa notes that he hopes to employ "45 units" in due course and we know that close to 60 individuals would be attached to CAAL in the years that followed (Busa 1965, 65). Busa's description of the workflow of CAAL, and the machines that were used there in 1961, evokes Winter's analysis of how the *Index Thomisticus* was at "the pivot point (or was the pivot point) between handmade scholarly tools and machine-made scholarly tools" (Winter 1999, 16). Something of the materiality of the millions of punched cards that the keypunch operators who worked for CAAL had produced by this point can be detected in his mention of the tedious work of moving punched cards between the various machines that were then in use. Busa writes that the project had also begun to use electronic computers for processing. At the time he wrote this, tests to see whether the programmes that had been written for an IBM 704 and IBM 7090 to search the texts that CAAL had captured were planned for August 1961.

References

Busa, R. 1980. The Annals of Humanities Computing: The Index Thomisticus. *Computers and the Humanities* 14(2): 83–90.

Busa, R. 1965. An inventory of fifteen million words. In *Literary Data Processing Conference Proceedings September 9,10,11 1964*, ed. Jess B. Bessinger, Stephen M. Parrish, and Harry F. Arader, 64–78. Armonk, New York: IBM Corporation.

Jones, S.E. 2016. Roberto Busa, S. J., and the Emergence of Humanities Computing: The Priest and the Punched Cards. New York; Oxon: Routledge.

Winter, T.N. 1999. Roberto Busa S.J. and the invention of the machine-generated concordance. *The Classical Bulletin* 75(1): 3–20.

I

Texts processed or in the process of being processed:

1) The main texts:
   a) Opera Omnia by Saint Thomas Aquinas.
   b) Manuscripts of the Dead Sea.
2) Texts connected to the main texts:
   a) Bonitz – *Index Aristotelicus*.
      Forcellini – *Lexicon totius latinitatis*.
      Thesaurus linguae latinae.
      Vulgata – Clementinae Editionis.

CAAL that is currently known dates to June 1956, when it was housed in *Casa Sironi*, which seems to have been a section of the home of the Sironi family in Gallarate, Varese (see Busa Archive #0025).

b) Zorell – *Lexicon Hebraicum.*

3) Occasional works:

Some of the "canti" by Dante.

Testi antichi Italiani (Ed. Ugolini).

S. Bernardi – *De diligendo Deo.*

E. Kant – *Prolegomena zu einer jeder Künftigen Metaphysik.*

Goethe – *Farbenlehre Bd. 3.*

Fabbri – *Veglia d'Armi.*

In all, we already have on record 3,500,000 words in 7 languages: Greek, Latin, Italian, German; 3 Semitic languages: Hebrew, Aramaic, Nabatean; and in 3 alphabets: Greek, Latin, Hebrew.

II

The results of the processing of each of these works:

1)

A. *Index cards*: text cards, word cards, form cards and lemma-cards.

B. *Lists:*

a) *lexical*:
laterculum formarum
conspectus lemmatum
rationarium verborum
index locorum
concordantiae

b) *statistical*:
index a contrario
index by length
index by frequency
index by environment (punctuation, neighbouring words)
index by special markers
index by endings (1, 2, 3 … n last positions)
index by initials (1, 2, 3 … n first positions)

All of these searches were first conducted on graphemes, coded at one character per column with a maximum of 3 perforations per column in 47 combinations.

2) Similar processing for similar results, carried out on texts in phonetic transcription: an automatic transcription having been made previously on the same texts: for an indefinite number of possible combinations, because each phoneme was represented by the combination of 2 columns.

III

Developing and refining methodologies:

1) Extrinsic methods, which concern the tools that are required for processing.
   A. Techniques for selection and training of staff. As a result, no more than 20% of candidates showed themselves capable of operating as tabulating machine operators.
   The Centre consists today of 30 units: 6 for philological work; 2 for the secretariat; 22 for the laboratory. We are setting out to achieve 40 to 45 units.
   B. "Cost accounting" and techniques for the continuous control of production, of time and of prices.
2) Intrinsic methods: that is, methodologies for the different aspects of processing
   A. planning,
   B. programming,
   C. the 3 parallel lines of operation:
   production,
   control,
   correction,
   D. Ancillary operations, often mundane, yet ... unforgiving:
   moving files of index cards to machines and vice versa
   "preparation" of lists
   compiling and archiving of programming descriptions and operations reports.
3) All these products and methodologies have been, and continue to be produced and explored with punch card machines, but we have begun to carry all the processing on electronic computers as well.
   A. Programmes are being prepared for IBM 704 and IBM 7090, the first tests of which are planned for August 1961, so that the same programmes are used to conduct lexical searches of any language (even non-Indo-European), of any text, and for any research requirement, including even statistical research and also comprising phonetic transcriptions, etc.
   B. We are searching for solutions to make the huge amount of electronically generated documents more manageable: for example, with the functionality of a "condensed file". Considering that the magnetic tape will preserve the entirety of the materials that are produced, it will be possible to distribute, at an accessible price, copies which are easy to handle or an extracted index. From this, one can determine, for example, some of the key words for an *Index locorum*, with only the references, accompanied by the *Rationarium*, and automatically print the texts on large index cards in folio format. From here, the researcher can extract, with reference to the Index locorum, those cards that contain the words that interest him.

IV

The international services of the Centre:
With funding from EURATOM:

1) to keep each of the centres at the international level informed about the other centres and about other ongoing work worldwide,
2) with a newsletter that will not be a separate journal, but, as it were, a press release to all interested journals,
3) to set in motion the standardization of terminology, of codes and of methodologies,
4) the dream of international centres: services by individuals or by the institutes to freely provide the first electronic versions of linguistic materials on cards or on tapes.

V Conclusion

We are conscious that all of us taking part in the colloquium are pioneers in the automation of lexical analysis. This illustrates a necessary role in the evolution of the "book", i.e. the material symbolization of human knowledge which is underway, and which is in the process of changing dimensions. Just as at the time of Gutenberg, the printed book took its place next to manuscripts, today, alongside the notebooks and printed books, the "electronic book" is taking its place. This, at least, is demonstrated by the fact that industrial production and defence feel obliged to finance research for the automation of information retrieval and mechanical translation, for which the automation of lexical analysis represents a preliminary, necessary step. This is a question of rendering a service to science and to human life with heavy and difficult work, because it requires an organization and systematization of one's own work, pushed to the most granular details.

The religious aspects of this specialization are signified in the words of the Old Testament: *Spiritus Domini replevit orbem terrarum et hoc quod* (there is a mistake here: it should be *hic qui*)[2] *continet omnia scientiam habet vocis* (the spirit of God prides himself on having the knowledge of one single human word as much as he prides himself on having the power to rule the universe!). Our linguistic science is therefore a participation in the linguistic science of God.

[2] [Editors' note] This comment in parenthesis appears in the original text and was presumably made by Busa. It is a puzzling comment because we have double checked the original biblical quotation and there does not, in fact, appear to be a mistake in the quotation. The phrase may be literally translated as: "The spirit of the Lord has filled the world, and this thing which contains all things has knowledge of the voice/word" (Wisdom 1.7)

Chapter 6 Linguistic Analysis in the Global Evolution of Information

Roberto Busa S.J.

First published as: Busa, R. 1962. L'Analisi linguistica nell'evoluzione mondiale dei mezzi d'informazione. In *Almanacco Letterario Bompiani 1962*, ed. Sergio Morando, 103–107. Milano: Bompiani.

Fig. 6.1. This photo was taken in Milan, at the conference entitled *Linguistica e Industria Oggi* (Linguistics and Industry Today) during the same year that the article below was published. To Busa's far left can be seen Leon Dostert, who led Georgetown's Institute of Languages and Linguistics and their work on machine translation. Busa would later write that he himself was the intermediary responsible for the installation of the Georgetown Machine Translation system in EURATOM, Italy in 1963/4 (see Busa 2004). Both locations are mentioned in the article below. (19/01/62, Busa archive #0467)

Editors' note

Busa begins this article by describing speech as tripartite in composition and he then turns to the increasing role of the computer in linguistic research. The sense of horror that this has occasioned among some humanists is powerfully evoked by the analogy of a bulldozer running amok on the meadow of humanism. He writes how the "men involved in automation" have asked humanists important questions whose answers are demanded by "economic facts". Humanists have been unable to answer them due to their inadequate knowledge of language. As a result, he argues that the contribution of the computer to

J. Nyhan and M. Passarotti (eds.), *One Origin of Digital Humanities*,
https://doi.org/10.1007/978-3-030-18313-4_5

humanistic research is not only to point to gaps in humanistic knowledge but to also offer the means of addressing them.

Busa then brainstorms uses for statistical linguistics derived from computationally-assisted research. For all his emphasis on the mystery of language he proposes that "A list of these percentages, extended to the speech and writings of thousands of pupils of different backgrounds—an extension made possible only by automation—would permit the identification of curves of normality that would be useful for further diagnosis of the human psyche at an age when a person is more amenable to educative influences". Given the post-World War II and Cold War context that he was writing in, it is unsettling that he suggests this without further reflection on the human implications of using technology and statistics in such a normative way.

From there Busa turns to the importance of his work and its impact on the global stage. He writes that his work was the first in a "sequence of events by which the automation of accounting caused the worldwide evolution of the means of information." From his earliest contributions he then traces four developments that hastened technological progress, including research on machine translation by research teams in Euratom of Ispra and the Georgetown Institute of Languages and Linguistics. This is one of the last articles in which he discusses machine translation in such a positive way.

References

Busa, R. 2004. Foreword: Perspectives on Digital Humanities. In *A Companion to Digital Humanities*, ed. Susan Schreibman, Ray Siemens, and John Unsworth, xvi–xii. Oxford: Blackwell publishing.

I

1. The phenomenon of language is greater than we are; it is one of the ingredients of that strange concoction that makes up each and every one of us. Indeed, the values of which we are such a fragile and marvellous tissue are in themselves much more diffuse and much greater than we ourselves. For example, our hands serve us for many things, both simple and complicated; they are, as it were, like waiters always attentive to our needs: we use them without paying much attention to them. If, however, we were to examine them and think about them a little, we should find in front of us a whole world to discover. Another mystery would be our capacity for aesthetic appreciation. By virtue of what "program" installed in the robot that we are do we feel so strongly, for example, the need for symmetry and the aversion to any mismatch of colour, line or sound? Yet, our hands are our power of expression; with gestures, facial expression, arts, and words we wound and heal, spill and raise, improve or worsen all that is around us. These worlds that are within us also need to be explored.

2. In our speech, which we use every day and about which we know so little, there are three levels: the conscious, the subconscious, and the unconscious. In that same area of our language that is enlightened by our perception and attention, one part, but not all, is amenable to control, in the English sense of the word: one part, that is, can be governed and thus also educated by us. It is more or less theoretically possible for a person from Milan to decide to get used to saying "*vada*" [go] in-

stead of the incorrect "*vadi*", to which for some reason of heredity he is so attached! There are other areas that refuse an organized control but are not completely resistant to a systematic survey: we are not able to change them, but we can in some measure take account of them. Other areas answer only to the subconscious or even the unconscious. For example, it is only with great subtlety that we come to realize that if we prefer some words to others, we do so because we are commanded by a subconscious desire to cut a fine figure, as it were, and we choose those words because of the greater value we attach to them, as ladies do when they use such words as "*genare*" [to embarrass] or "*flattare*".[1] Some people choose words on the basis of their capacity to define things, others solely for their aesthetic effect, whether phonetic or semantic, or for correlation and rhythm. Yet at a deeper level, grammatical and syntactic structures seem to sprout from the unconscious roots that have nurtured human life in its vital evolution in the universe wherein it has been active for such a short time just like an amoeba in its culture medium![2] The roots of language are to be found in the human behaviour zones that are inaccessible to education and self-control, because they are programmed and operated exclusively by what lies at the roots of our physiology and of the regrettable and inevitable mixture of physiology and pathology.

3. We cannot, then, know everything about our speech. Of the great or small amount that we can know, not all of it allows itself to be influenced by our aggressive ambition to do both with us and through us "what we want". Nevertheless, why could we not leave uncultivated that flower-bed which we could dig if we wanted? Why should we want to divert some current of water from the river of our words to send it through sluices? Precisely for the same reason that we try to control water. Speech is the principal potential energy that man has and it is therefore distributed economically. Ideas are force only when they can be spoken and written; they have no other way of taking hold outside the individual that has them.

4. Aristotle, therefore, purposefully set about looking into language, and in its folds, he discovered metaphysics. Aristotle was an encyclopaedic and practical collector of facts, among which the following fact was one of the most remarkable: moving on the path of Aristotle is to feel oneself catapulted into a steep climb from the path that he investigated, step by step. Even the good and gentle St Thomas Aquinas stood and admired Aristotle, wondering with his head in the clouds at the power with which a pagan from this earth had been able to enter into heaven. But Philo, who knew the Old Testament, and Christian Theology, and started from the examination of the "word" penetrated much further into the heavens, surpassing by far the great achievement of Aristotle and Plato. Amongst other words they caught a glimpse of the dazzle of the Logos, the Verbum. And there has never been an absolute idealist that has succeeded in attempting as much as

---

1 [Editors' note] These words have become rare and old-fashioned. *Genare* means to embarrass; *flattare* is a gambling word.

2 [Editors' note] Here, the Italian text says "*brodo di cultura*", lit. "culture broth". It is an ironic expression for "manifestation of great culture".

the Christian philosopher did in his deep understanding of the expressive meaning of the "verbum mentis" and of the consequent reverberation of mutual love within that absolute thought which is the fire of consistency, life and imagination; the director who at the same time is the voltaic arc that projects onto the dark screen of nothing that succession of images that are us, the world and history.

5. In social life, grammar and logical analysis have educated for so many centuries that indefinable *je ne sais quoi* which we call humanity and humanism: i.e. that touch of enjoyment of beauty, sense of harmony, that appreciation of formal values, by which even at the *Politecnico* [an Italian academic institution] those who come from a *Liceo Classico* [roughly equivalent to a secondary school] and those from other schools can be distinguished. Rhetoric influenced the way in which we express ourselves. Aristotle's dictum was "*signum scientis est posse docere*": our knowledge of a subject is fully developed when we are able to pass it on. We have all realized by experience that when a teacher takes two hours to get us to understand something, it is because he himself still does not understand it perfectly, and that for him to understand it perfectly he had only to study it properly before talking about it. How often have we seen the infinite wisdom of the saying that there is an enormous gap between being right and knowing how to prove it! Today we are often preoccupied merely with swallowing ideas, as if man were a general store, or as if he were composed of nothing else but memory. Man, however, is especially and designedly an organizing and resourceful power, and in his education, he should not be treated as though he were a rucksack to fill with what is needed for a camping holiday; his mechanism should be polished, oiled and well run in, like that of a machine tool that can then work for a long time on any sort of material. Would it be a mistake, then, to think that it would be worth knowing half of what we know, if thereby we could explain better the little that we do know? For this reason, greater care was formerly taken with our means of expression. One was educated so to organize one's thoughts as to select the right combination of words for what one wished to achieve: i.e. to think how to speak before speaking. Yet even Manzoni used to say that this one thing "thinking before speaking" is in itself alone so difficult that we too come close to apologizing for the number of times we give ourselves up to speaking at random.

6. Ever so gradually the universal laws of ageing, which eat away at institutions such as man and nature, have weakened the grip of logical analysis and rhetoric. The power of decadence has extended its tyranny to such an extent that today we go to drama school only for a career in acting and not to prepare ourselves for delivering our lines in the comedies and tragedies of life. If it has been written that the divine nature of the Gospel is shown only by its survival in Sunday sermons, if, that is, you middle-class people find that we priests are often so careless in our prayers, it is because we, like you, are the children of our time. The sunset appears purple also because in Italy Latin is treated like an old grandfather for whom we wish another hundred years of life, while our subconscious registers that at the bottom of us we do not feel a truly infinite horror to think that soon we shall be without Latin.

7. At this point a nightmare intervened, technology triumphant with its latest creation: automation. People shuddered, considering it a crude, hard bulldozer that goes roaring ahead, crushing and shredding flowers, amongst which, a delicate and gentle victim, is humanism. Tomorrow is already upon us. The future has already begun: the lava-flow is spreading out and burning the green sides of the mountain. Inside the monster's operations room are men, encapsulated amidst gauges, cloches, warning lights and dials. Perhaps at the start they were not even aware of the elegiac wailings and lamentations of the "humanists". In fact, they are satisfied with working. They claim to offer a service of public utility, because they consider that without them industry and commerce would no longer be able to answer to man's needs. Then, however, not yet ten years ago, the men involved in automation began to make their head lean out from its cabin in the tower of electronics and ask philologists and grammarians, who were busy in the fields selecting the choicest flowers, questions such as these: Please, how many verbs are there in Russian that are active and transitive, and how many that are active and intransitive? How many are there in English? What is the greatest number of initial and final letters that coincides with the greatest number of words? Which words or linguistic situations are found within a radius of *n* words, only when and always when "*faccia*" means "face", and which others only and always when "*faccia*" is a form of the verb "*fare*" ("to do/make")? Please, would you arrange all the words in the dictionary according to the various morphological and grammatical categories? Would you please tell me which words may be omitted, and when, so as to shorten a text without any detriment to its meaning? Can you tell me precisely the characteristic environment of certain semantic categories that are neither morphological, syntactic nor structural? What happened was sensational: a machine made us realize that no humanist has such command of his own language as to be able to answer such questions. A machine, the skivvy of banal commerce and drudging industry, has revealed that there is still too little humanism of the serious and systematic type. Economic facts today demand a qualitative increase of grammatical and lexical sciences as one of the necessary conditions of their vital development. Yet they also offer the possibility, and this is neither petty revenge nor small satisfaction.

II

8. The Centre at Gallarate is still today the one in the whole world that has put the greatest number of words on cards: there are to date about four million, and the number is increasing. Seven languages are involved (those of Aristotle, Ancient Italians, Dante, Kant, Goethe, Hebrew Texts from the Dead Sea, Fabbri, etc.) and three alphabets, Latin, Greek, and Hebrew. However, when in 1946 I began to think seriously about the verbal indices of the thirteen million words of St Thomas Aquinas, and when later, in 1949, I began the first experiments with IBM, and when still later, in 1951 I published the first results, not only was I the first and

only one in the world to venture to saddle the flying horse with lexicology, but I was also unaware of the historic moment in which that happened to me. If it did not come to me, the idea certainly would have come to someone else, and perhaps one day it may be known that it came to someone before me, to whom nobody at the time had paid any attention. If it is permissible to speak of merit, this, if at all, would be found in the great patience needed gradually to resolve all the unforeseen difficulties that one meets in transforming an idea into a mature and practical methodology that can be applied, so to speak, to a production line. Of the famous dictum "genius is one per cent inspiration, ninety-nine percent perspiration", the only word that does not apply to me is the first. Who could then have imagined that punch-card machines would today be considered antiquated, and that we should have seen the evolution, or rather, the transformation of electronic calculators from ones with superficial memories coated with ferrous oxide to those operating with a ferrite ring network, and finally to the cryogenic ones, which are very thin films superimposed upon each other like the pages of a book, and which can be used at nearly zero temperatures? I certainly never imagined that the "stretch", which was constructed for nuclear research, would have possessed a memory just short of two billion positions, in which the whole of the *Treccani Encyclopaedia* could wallow like a baby in a big bed, and another memory of one and a half million positions that has an access speed of some hundredths of a millionth of a second. Above all, I was unaware of the fact that I was placed in the sequence of events by which the automation of accounting caused the worldwide evolution of the means of information.

9. I can reduce to four phases the movement that after 1945 acquired the acceleration of an avalanche. First phase—The development of communications and organizing techniques allowed the great enlargement of firms that were able to operate world-wide. Equally rapid was the increase of the reciprocal influence of the markets, and between politics and the market. With such a development it has become indispensable for a manager to be able to review a great number of particulars and quickly synthesize them, and at the same time to check and perhaps modify the great masses of small and extensive peripheral phenomena. Calculators answered this need and provided economics with industrial and commercial accounting. They eventually were able to perform up to a million multiplications and divisions every second, and to print the results of their calculations at a rate of 60,000 words, and 300,000 numbers per hour.

10. Second Phase—Industry, whose development is greatly increased by the demands of "defence", and the parallel deepening of the relationship between industrial production and scientific research have imposed the automation of scientific calculation. Euratom, for example, has felt obliged to acquire for its own *Centro* of Ispra the IBM 7090 calculator, which costs about three million dollars, or about two thousand million lire.

11. Third Phase—The activities of production, trade, and defence demand the automation of "information retrieval", which I would translate as an opportune sys-

tem for the tracing of useful knowledge. The amount of scientific publications, already enormous, is continuously increasing. The USA to-day has an average of 40,000 new patents each year. On the other hand, the acceleration of scientific evolution is such that publications about nuclear physics after two years belong merely to the history of physics. However, as regards computer technology, the most up-to-date information represents the crest of a wave for slightly more than six months. Now imagine that the missile industry needed to know the behaviour of certain materials in certain new situations. How long would it take to sift through everything that is known in all the relative sciences until one found what was needed? Analytical indices would not be any use, because by definition what is being sought is not common knowledge; neither would bibliography be sufficient, for this deal merely with titles, while what one needs, for the reason given above, is to rummage through the contents of what has been printed. If anything, abstracts would be useful. But try to read them all and you will tell me when you have finished that it is too late. How, then, can one keep up with all the publications throughout all the world at almost the same time as they appear? I think that it has happened to DDT to be discovered for the first time two or three times successively! Therefore, it is necessary to condense the greatest possible amount of scientific information in such a way that everything needed for new research can be identified in the shortest possible time. Automation aims to reach that point.

12. It has been channelled into the following areas: new types of symbolization of knowledge, in other words, magnetic imprints of alphabets; how to transcribe and recopy with these alphabets, which only the machine can read, the content of which has been printed in ink on paper (much hard work is being done so that this will be possible by photographic and phonetic means [at the time of translation this would be equivalent to work in OCR and speech recognition]); how to condense this (to summarize it, reduce it to a telegraphic style, to abbreviate the words); how to classify it, and how to search for it. A chapter of this effort is represented by machine translation. I do not mean the science fiction fantasy of machine translation of a literary or philosophical text, but the technique of machine translation of contemporary publications on the same subject, in sciences unified as they are to-day, and for that reason thought out and expressed in the same manner and with a vocabulary whose only differences are those of the two languages. To the problems above this technique adds that of "how" to that of "which", according to distinctive contexts of a word, its grammatical and logical function can be identified by a machine, and, in the case of polysemy, its meaning in a particular context. There is also the problem of how a machine can transport the syntax of one language into that of another. One year ago, the University of Georgetown, Washington DC, opened a centre at Frankfurt am Main where thirty people continuously punch Russian scientific publications, which are then translated into English by the 704 calculator.

13. Fourth Phase—Automation of the treatment of information requires the automation of the compilation of indices, concordances, and of all the possible types of statistics of linguistic facts. If you go to the Euratom of Ispra and visit the

CETIS group, or go to the Georgetown Institute of Languages and Linguistics in Washington, you will realize that a new lexicology and new linguistics into techniques for the treatment of information are developing amongst the researchers. This lexicology and linguistics are more systematic, more exhaustive, more widely useful, and, I am emboldened to say, more humanistic than the traditional ones in use up to now. It will not be long before the tenor voices of philologists will sing the praises of automation in the gardens of humanism, accompanied by the baritone comments of mathematicians.

III

14. Indeed, there are mathematical formulae even within our speech, which is the greatest expression of our freedom, personality and capriciousness. One cannot even "speak as one pleases" without obeying certain laws. If you were to abandon yourselves to the pleasure of abandoning certain boundaries and, from the great sea of combinations that are arithmetically possible between the elements of our vocabulary, come out with certain sequences of words that are unusual within those boundaries, be well assured that they would lock you up somewhere and would subject you to sleep treatment. But it is not just in this sense that there are rules in speaking. Numbers underpin language (what joy Pythagoras would have were he still alive!), just as proportions of measurement and connections of relations constitute the skeleton of shape and beauty. Linguistic statistics, which our Davanzati used a few hundred years ago, are more encouraged because number still holds sway among the bases of ideas and logic, as is demonstrated by symbolic logic and algebra of propositions, just as it belongs to the substance of the foundation and fount of being, as Catholic Trinitarian theology reveals. Then, since language can be translated into combinatorial terms of a great mass of small elements, i.e. since it is a lattice of repetitions and frequencies, its mathematics are not just deterministic, but rather, and indeed more so, those of probability and chance: wonderful mathematics nearer to the mystery of God, spirit, and art. The polymath Johann Joachim Becher, who died in 1682, and who invented the theory of phlogiston, deserves everlasting gratitude from his mother Germany for having taught her how to get alcohol even from potatoes. He was a man of vast empirical interests and may be called the precursor of the numerical codification of words.

In his *Character pro notitia linguarum universali*, Frankfurt 1661, he proclaimed that with just one language every other language can be understood, on the condition that every concept be expressed with a number or a corresponding hieroglyph. This is exactly what is needed, and is still lacking to a large extent, although much work is being done in this area, so that any computer, digital or analogue, can serve us as a faithful and discreet translator: the calculator that the Germans call *Hochgeschwindkeittrottel*, high velocity cretin!

15. The following few examples will show the usefulness of statistics of linguistic factors, statistics as vast as the incredible possibilities of automation allows them to be. At Gallarate, Professors Tagliavini and Croatto of the University of Padua carried out the automatic phonetic transcription of a text of Fabbri of about 20,000 words. They then went on to list the phonemes and tri-phonemes of spoken Italian. The thesis in which A. Zampolli presented their conclusions created a great uproar, because the most frequent tri-phonemes at last were recognized, i.e. those which concur in forming the greatest number of words. With these, work has now begun on the re-education of the deaf and dumb, so that they do not experience the annoyance that we had (do you remember?) when as children we were crammed full with exceptions in French—*hibou, genou, caillou...email, epouvantail...*—with the result that today each of us knows perfectly the words that we never use and makes mistakes with everyday ones.

The proportion of the use of nouns, verbs, adjectives, prepositions etc. oscillates around fixed points, which vary, however, according to age, sex, temperament etc. A list of these percentages, extended to the speech and writings of thousands of pupils of different backgrounds—an extension made possible only by automation—would permit the identification of curves of normality that would be useful for further diagnosis of the human psyche at an age when a person is more amenable to educative influences.

This summer the newspapers have reported worldwide the conclusion of the metrical census of the Iliad undertaken at New York by James McDonough. The conclusion was that the Iliad was the work of one single author. I started this young American scholar off a few years ago. He began to punch onto cards the whole quantity of syllables of all the verses. When the whole of the Iliad had been transcribed in this way, a computer showed the rhythms and the proportions of the use of the various metrics. If one were to do that without a computer, apart from the length of time that it would take, one would still have to do it right from the beginning in the same way. Today, if you cannot believe the conclusions, you can check all the calculations in a few minutes starting from the cards at the beginning.

The chronology of Plato's works was also established, and this is universally accepted today precisely because of the statistics of the stylistic elements, although this was carried out without the help of automatic machines. With a similar procedure the problem of authorship of various works could be tackled, for example those of Shakespeare or Marlowe, precisely because there always exists in anyone's style permanent personal traces of his characteristics as well as of his own fingerprints. This is obvious if one considers that whatever we express, it is through ourselves that we express it.

There are studies on the sharpness of stress accents. Words with the stress accent on *i* or *e* express lofty and intense sentiments; those with the stress on *o* or *u* express depressing ones, and those with the accent on *a* express neutral ones. A passage from *I Promessi Sposi* and its French translation was analysed. The fluctuation of the sharpness of the respective stress accents was shown with curves on a piece of graph paper. On another piece of graph paper was the curve showing the succession of the various levels of emotion expressed by the words of the text. The result was that the curve showing the movement of the sharpness of the stresses in the original coincided with the curve showing the sentiment. This was not so in the French translation; here the phonetic rhythm of the stresses neither welled up unconsciously from inner inspiration, nor was it considered as an element to be "translated".

IV

16. At the time of Gutenberg, the printed book took its place alongside manuscripts that were in the form of notebooks and registers. Books and manuscripts will remain, and currently the "magnetic book" takes its place by their side. For the storage of human knowledge this represents a real change of dimension. But it has not merely to do with quantity and speed; it is also a matter of quality. If indeed it is true that the language of electronic calculators will in all probability put a stop to attempts at universal artificial languages, it is also true that the inductive interpretation of the phenomenon of language with the help of probability formulae promises (i.e. the evolution of language towards new specifications and at the same time its involution or entropy towards gradual loss of semanticity: here too are laws of the mixing of life and death), insofar as it makes automation possible, to restart the cycle of linguistic and grammatical awareness with greater depth, methodicalness and documentation.

17. One of the signs that we are here too at a crossroads is the fact that there are in the world about 200 centres dealing with this collapse of the tower of Babel, in order to safeguard the unifying physiology of communication in language and to hold it back from the pathology of barrier and separation. Among these centres there are now a dozen, after the one at Gallarate, that deal with pure lexical analysis. The others apply it to techniques of information and machine translation. The development of the automation of language is in fact triangular. Another sign can be seen in the fact that institutions such as ministries of commerce and defence and others—USA, URSS, Nato, Euratom etc.—have financed it for some years for this purpose. In France, Holland, Israel, Czechoslovakia there are gigantic projects of computerization under way—120,000,000 cards for the *Trésor* of the French language have been mentioned—from which to get the material for the compilation of dictionaries of the national languages. The card index of the *Index Thomisticus* too could be defined as the first Thesaurus of scientific language of our Middle Ages.

18. Domenico De Domenichi, a Venetian, "*de ordine plebejo*", became the vicar of Pope Sixtus IV. In a preface to an incunabulum printed at Venice in 1480, he commented as follows on the recent invention of printing: "*Placuit autem clementissimo Deo his nostris temporibus novam artem docere homines*" [in these days it has pleased our most merciful God to teach mankind a new art]. He goes on to

report the amazing news that three men have been able to print three hundred copies of the book in three months: "*ad quae tota eorum vita haud quamquam sufficeret si cum digitis aut calamo aut penna scribenda forent*" [the whole of their lives would not be sufficient if they had to write the copies with their fingers using reed pens or quill pens]. He concludes by saying "*si quid in me est auctoritatis etiam admoneo: ne tanta Dei beneficentia abutantur*" [if there is any authority in me, I warn them not to misuse such a great a gift from God]. What should we say today?

Chapter 7 Latin as a Suitable Computer Language for Science

Roberto Busa S.J.

First published as: Busa, R. 1964. De Linguae Latinae Opportunitate et usu ad Scientias per automationem euulgandas. In *3me Congrès Int. pour le Latin Vivant, Strasbourg, 2–4 Septembre 1963*, 119–121. Avignon: Aubanel.

Fig. 7.1. The attendees at the *3me Congrès Int. pour le Latin Vivant* (third international conference for living Latin) in Strasbourg where Busa presented this paper, 02/09/63. (Busa Archive #0536).

Editors' note

In this lecture Busa described the shape of the *Index Thomisticus* as it was foreseen in 1963/4. At that time, it was intended to comprise four parts: the first part dealing with texts of Aquinas; the second with his commentaries; the third with Forcellini's *Lexicon totius Latinitatis* and the *Thesaurus Linguae Latinae;* and the fourth with "selected works useful for obtaining a method", including the *Qumran* and Goethe's *Farben Lehre.* As later articles will show, some changes to these divisions would be made in the following years. Aside from discussing some of the problems of linguistic disambiguation that he was facing, especially with homography, Busa is otherwise concerned to celebrate the Latin language in the article below. He proposes that it should become the *lingua franca* of science once more and that this would offer "an easier and cheaper alternative to

J. Nyhan and M. Passarotti (eds.), *One Origin of Digital Humanities*,
https://doi.org/10.1007/978-3-030-18313-4_6

computer translation". This is presumably an implicit reference to the setting up of the ALPAC committee in April 1964 to examine the state of the art of machine translation (see Alpac 1966).

Works cited

ALPAC 1966. Language and Machines: Computers in Translation and Linguistics. A Report by the Automatic Language Processing Advisory Committee, Washington, DC. Publication 1416.

1. The Reverend and most distinguished Don Paul Poupard, after he had reached the height of his career in this faculty of arts, in 1961 published a work about Louis Bautain, Professor here in the University of Strasbourg. This excellent example of scientific method took many thousands of hours of hard work to produce. In it he raised the question of the power of religion in languages and of their religious origin (Paul Poupard: *L'Abbe Louis Bautain*, Paris, 1961). This was not because Bautain was a traditionalist, so to speak, but because he recognized the divine origin of languages (pp. 165 and 294 s.). The reason that many of us priests have come here, attracted by the fascination of your Latin studies, is not only because we like to join you in ministering to human beings who communicate with each other, but also because we know that souls turn to God in this way and thus we serve God.

2. What I am about to say is divided into two sections, each of which is further subdivided into three. The first section deals with why I am here addressing you. The other, however, is, as it were, a testimony to those who devote their energies to computers and who admire your splendid hard work.

I

3. First of all, you should know that I am not a Latin scholar but an engine driver. I am in charge of 40 assistants and 30 computers at the "*Centro Automazione Analisi Linguistica*" at Gallarate near Milan. Within six months I must have 70 assistants, yet we must finish the whole *Index Thomisticus* with the IBM computers by the end of 1965.

The whole *Index Thomisticus* is in four parts, the first of which deals with the *opera omnia* of St Thomas not excluding the *dubia*; the second with his commentaries (*Biblia Sacra Latine, Aristotelis Opera Graece, Petri Lombardi in 4 LL. Sent. Latine*, etc.); the third with the *Forcellini Lexicon* and the *Thesaurus Linguae Latinae*; the fourth with some selected works useful for obtaining a method (Texts of the *Qumran*, Kant's *Prolegomena*, Goethe's *Farbenlehre* b.3, St Bernard's *De Diligendo Deo*, and some others): all these texts together contain about 15,000,000 words in eight languages and three alphabets, Latin, Greek, and Hebrew. Concordances of every word without exception in all these works are being prepared: these will give every usage and context of each word in about one hundred characters or about 12 to 15 words. Four other indices will accompany the concordances as an index and guide to them, since their tables are compendious and purely statistical. These indices are: survey of the lemmas (*conspectus lemmatum*), list of forms (*laterculum formarum*), table of words (*rationarium uerborum*), overview of homographs (*prospectus homographorum*).

4. The use of computers has sucked us down into a bottomless whirlpool. Since half of the words in Latin are homographs, and since it is well-nigh impossible to consider every single one individually, we were compelled to attempt first of all to classify the types or grades of Latin homography. We then compare the single forms of the words taken from those texts with those single lemmas of Forcellini with which they might have some graphical proximity. Lastly, we choose the homographs that must be distinguished from one another. For this reason, I beg you, out of the kindness of your hearts to give us any help, advice, and support that you have immediately to hand for dealing with Latin homography, or any other Latin lexica that you think we should examine electronically, or any lists of forms or endings of Latin words, both those that are found in extant works and those that are theoretically possible.

5. However, let me give a word of warning that applies not merely to homographs but to every subject: electronic machines can work only if programmed in a systematic manner. We divided the writings that we are working on into thirty operative units, each consisting of about five hundred thousand words. We give a complete index and concordances of every operative unit, of every word without exception, together with evidence of repetition and combination. All these must always be reduced to a system.

6. Up to now no grammar book or lexicon has presented the elements of Latin in such a complete, consistent, and systematic way as a computer is programmed to do. Far from diminishing humanism in any way, computers actually promote our humanism to the perfection of a scientific method. Please make a point of saying this to anyone who fears that the use of computers will impair our mental faculties. Computers invite us to wider, deeper, and more systematic research. (Perhaps the only thing to be afraid of is that one might be overwhelmed by too much exercise of the intelligence). Not only do computers invite us to wider, deeper, and more systematic research, they also make it possible.

II

7. In order to make all this more understandable, let me tell you some of the history behind it, in particular about a conference on developments in communications that was convened by the military commanders of the states belonging to NATO. The conference was held in Washington and it finished on the 17th of August this year. I was sent to the conference by the Italian government and I have just come

from there to you pacifist soldiers of Latin, which you rightly assert to be the best vehicle of communication. From this, as it were, universal Academy of computing, I will offer for us three pieces of evidence, with some preamble.

When in 1949 I proposed the use of computers for a wider, better, and more accurate exploration of speech and texts, I was the only person in the world who did this. Today there are about fifty institutes that use computers to research into words: of these more than fifteen are engaged in preparing computerized concordances for the study of words, the rest deal with applied linguistics. My three pieces of evidence are as follows.

8. Very soon a huge number of texts will be arranged into concordances as evidence and sources for the making of an historical dictionary of modern languages. For this purpose, work is now being done at Nancy in preparing a thesaurus of the French language (up to 120,000,000 words!), *idem* at Leiden in South Holland for the Dutch language, *idem* at Jerusalem and Prague for Hebrew and Czech. At New York computerized concordances of the *Jerusalem Talmud* are being prepared, and at Bonn concordances of the works of Kant.

9. Linguistics, providing tools for people to communicate with each other throughout the world, runs up against the incredible difficulties of computer translation. For, although more money than one could ever imagine has been assigned to research into computer translation by military and technical institutions, almost nothing has been achieved, so that now there is much scepticism about this matter. However, experiments on artificial texts of a few words have yielded an excellent result. Yet for computer translation of a very great number of natural texts to be an easy, common, and a daily occurrence, so many "parameters" are still lacking that it will take many years and much hard work before we can hope for any result.

10. Day by day more and more institutes endeavour to reduce scientifically into a compendium even the smallest elements of every language: the "Basic language" which is rapidly studied. This is so that the greatest number of people may easily and swiftly learn the essential and sufficient minimum of every language.

11. From these three facts I will elicit some more observations. If anyone is setting about preparing an historical lexicon of scientific Latin from the XIIIth to the XIXth century, he is to be applauded wholeheartedly and must certainly be given all possible assistance.

11.1. An historical lexicon of any modern European language will never be completed unless a lexicon of scientific Latin is first produced. For without any doubt the greatest part of scientific words in modern languages comes from Latin, the universal language of science.

11.2. An historical lexicon of scientific Latin, in conformity with the fourth time dimension, greatly extends human communication: indeed, this alone allows us to communicate with the fathers of the sciences that we have today and who were pre-eminent in past ages.

11.3. For a lexicon of scientific Latin to include not only the scientific Latin words of our time—as used by those few who, like a few scattered swimmers in a vast whirlpool, write about the sciences in Latin—but also those words used in those ages when all writers wrote about the sciences in Latin, the following consideration should not be neglected: the Latin scientific language should not be exhibited as a new kind of language, but as a language that already exists, in fact, physically mature and not yet senescent, which human beings have briefly forgotten, as if swept away by the intoxication of new scientific inventions. If aristocratic genealogy and pedigree still have any sway, no scientific language is more noble than Latin!

11.4. I believe you would all agree that a list of the main works published in Latin in previous centuries about the individual sciences should precede an historical lexicon of scientific Latin.

11.5. Let us take as an example the bringing back to life of the Hebrew language. As we all know, this language, which already in antiquity was thought to be extinct and which in the Middle Ages was used almost only by Rabbis, has now come to life again. Indeed, it is capable of expressing even the most modern concepts, and that is greatly to the credit of the people of Israel.

12. Since fortune favours the brave, why do you not propose the use of Latin as an easier and cheaper alternative to computer translation? Why do you not confidently and solemnly hold up to governments, which control all commerce, industry, and defence, the Latin language as an excellent solution, and one ready to hand? Russian scientists themselves have proposed this recently! At least let Latin and computer translation fight it out as rivals and competitors, and let them be weighed in the balance: would not Latin win, if only we wanted?

13. Lastly it should be established from the very evidence of the facts that the universal use of scientific Latin would cost less than computer translation, if we consider the following two points. First: worldwide communication is as immediate as the publication of a written work.

14. The second follows from the third piece of evidence given above. Naturally today support is given to teaching languages to a great number of people. Therefore, if a basic scientific Latin were prepared from the historical lexicon, it is certain that all adults who study sciences could master it in one year at the most.

15. To the three pieces of evidence I will add a fourth. I offer the following opinion, and submit it to your judgement: at the present time God's Providence has ordained such opportunities for Latin to again become the universal means of scientific communication that only one thing appears to be lacking: namely, that you, or rather, we all, work hard towards it. Success depends on what we do. If, instead of dreaming, we are prepared to act, work, produce, achieve, construct and cooperate, by God's good grace, Latin will again be the international language of science.

16. Finally, I would ask you whether you agree with me when I propose:

1. That a programme or agenda be prepared by our valiant guardians of Latin;
2. That they, like Gideon in the Old Testament, ask us which of us is prepared to shoulder at least part of the work himself.

Chapter 8 Cybernetics and the Possibilities of a New Human Being

Roberto Busa S.J.

First published as: Busa, R. 1966. La Cibernetica e le possibilità dell'uomo nuovo, *Il Fuoco* n. 3 (Maggio-Giugno), 19–33. Roma.

Editors' note

In this text, Busa discusses some of the consequences that cybernetics could have on "human beings of tomorrow". He lists a number of definitions of cybernetics, including "any form of automation". Though he notes that definitions of automation can also be elusive, it is with the idea of cybernetics as automation writ large that he frames his argument. Busa supposes that cybernetics will have two far-reaching consequences for society: it will occasion a quantitative turn in approaches to problem-solving and herald the advent of an electronic library. He also suggests, more by allusion than direct reference, that cybernetics might be used to control society. In an article in the Catholic media that was written about his work, and which draws extensively on the text below, Busa discusses the potential societal impacts of cybernetics more freely (see O'Grady 1966)

The main concern of Busa in the article below is the potential consequences of cybernetics for belief in God. The cold war context in which Busa was writing seems to be underlined by his use of a quotation from Leonid Il'ichev, the "secretary of the Central Committee of the Russian Communist Party" to exemplify the view that cybernetics favours atheism. Busa rejects the view that cybernetics makes a case for aetheism or materialism by claiming that "if there is organization, there is a program, and if there is a program, there is a programmer [i.e. God]". From here, he performs one of his favourite pirouettes: just as we have seen him argue that literary data processing requires increased spiritual industriousness, here he argues that cybernetic research opens up ways of detecting God's existence. Again, we see how Busa's Jesuit training shaped his view of scholarship as he argues that this should be the ultimate aim of (cybernetic) research.

References

Desmond O'Grady. The Jesuit who punches cards. *U.S. Catholic* 1966, 31–37

I will proceed as do those who cross a river full of water by jumping from stone to stone.

1. I shall begin by making five points about the meaning of the term "cybernetics".

Cybernetics has been called the science of comparing command systems between the animal nervous system, on the one hand, and on the other, electronic calculation and systems of telecommunication. In this sense cybernetics is equivalent to a science of controlling. Cybernetics also indicates any use of automatic calculation. You have heard tell of cybernetic medicine or of the cybernetics of railway transport; if these words have any meaning, they mean above all the use of mathematical symbolization, and thus of electronic calculation in the study of these sciences and techniques.

J. Nyhan and M. Passarotti (eds.), *One Origin of Digital Humanities*,
https://doi.org/10.1007/978-3-030-18313-4_7

On a wider level, cybernetics means any sort of automation. It can be difficult to say what automation is. The following is the most apposite definition that I have found: automation is the technique of *feed-back*, or rather of retroaction or forced feeding i.e. when the operative process is modified by its own result. This is called the second industrial revolution: not merely is the execution of the work entrusted to the machine, but also the very choice of how to perform it. This meaning of cybernetics is often extended by analogy to other sectors. I remember that about eighteen months ago the editorial of an English magazine about electronics, "Control", described Khrushchev's attitude as cybernetic. It simply meant that he did not carry out rigid and univocal realizations of pre-arranged programs but modified his own behaviour on the basis of its results.

Cybernetics is the science and the technique of the "automaton", the robot. If you were to ask me if I always and everywhere see clear boundaries between science and science fiction with regard to robots, it is only with some mental reservation that I should be able to avoid saying no![1] The fifth and last meaning of cybernetics is when the person who mentions it does not make clear what he means by it! It is a word that bewitches and often carries the consequences of this with it!

2. It is necessary now to examine, or better, to give a summary of cybernetics with regard to human beings of tomorrow. I shall limit myself to describing those developments that will affect the majority; therefore, I will say nothing about those developments that will remain as rare and advanced points. I shall also touch on those social phenomena that are most likely to come about in the near future. You will have immediately understood what technical developments are predictable on the basis of automation. In places of scientific research, cybernetics will bring two features to human beings of the future.

Firstly, mathematical formulation will be applied ever increasingly to every problem: look how economics has become econometrics, and how statistics has entered religious sociology and has assumed the role of an auxiliary to pastoral

---

[1] On the subject of "machines that recognize and describe", "machines that think" and of the "secret of the making of man", it seems best to place the argument on three levels. Above all there is the good solid learning of someone in a centre for psychological analysis who applies the resources of operational calculation both to the examination of the electro-neuro-physiological support of knowing in particular, and to that of the same process of knowing in general. The perspectives of the method emphasized by Prof. S. Ceccato are given in his recent address at the Wiener Memorial Meeting in Genoa (26th–30th October 1965) "The mechanization of thought processes and of language". The "mechanical" models of the cognitive processes that it is hoped to obtain find their first useful and practical applications in the techniques with which man tries to develop instruments that allow him to widen and deepen his zone of operative and communicative influence.

I call the second level that in which everything is embellished with the stylistic techniques that are characteristic of journalism, whose function is to stimulate the attention of the mass of readers who have been made careless by the shock of the paradox.

But on the third level I place, at the sign of science-fiction, all those who, in affirmation of the type "artificial reconstruction of a human being", would not see at the most the function and methodological values of the "limit" but the formulation of a concrete and realizable objective.

theology under the guise of algebraic formulation! Amongst all the resources of science mathematical calculation is one of the factors that accelerate its development.

Secondly, automation will bring to human beings of the future an "electronic library", to coin a phrase. We are living in a period similar to that of Gutenberg, when the printed page started to rival manuscripts. Alongside the manuscripts that now survive, and the printed pages that will survive, electronic memory is taking its place: the product of scientific research must, in the very nature of things, be condensed so that it will be possible to sift in a fraction of a second to find what is wanted with the certainty of having collected everything.

The social consequences of cybernetics are strictly linked to the practical diffusion of electronic calculation. Its characteristic is the ability to control an ever more detailed and ever more extensive periphery of vision, by representing it in the form of approximate figures on the organizer's dashboard. It is enough to mention teleprocessing and real-time calculation. It involves a direction of development in the sense of ever wider organizations at the base and thus at the centre of an ever-diminishing number of vertices. In other words, cybernetics will develop for mankind everything that is to do with organization and control, and with all the consequences.[2] I will mention just one among the consolations: the fact that to-day scientific research is no longer confined to national or international blocks, but is progressively freeing itself and escaping from every political barrier.

3. And yet human beings of tomorrow will need not just scientific and technical information; they will still need goodness! What goodness can the development of these sciences and techniques bring to us? Goodness does not have a specific weight nor is it to be found in Mendeleev's periodic table. But if we cannot manage without it in our lives, how can we say to each other that it does not exist? It is only when we suffer its lack that we appreciate its worth: it is something without which life should not be lived.

If it is true that goodness is connected with religion, what is the influence of cybernetics on religion? By religion first and foremost I mean the surrender by human beings to the discretion of the mystery of God. God is certain, just as death is certain, just as everything that is obscure in and around our life is certain. I repeat, for mankind religion is to surrender oneself to the mystery of Him. Cybernetics sets off in two opposite directions with an infinite range of intermediate situations.

Here is an extract from the report of Leonid Il'ichev, secretary of the Central Committee of the Russian Communist Party and also in charge of ideology, which

---

[2] Prof. G. Salvini recently underlined this for scientific research in itself, in "Metodi e mezzi del fisico, da Galileo ad oggi" v. "*De Homine*" n. 13–14, Ist. di Filosofia, Univers. di Roma, 1965 p. 44.

was read during an enlarged session of the Ideological Commission on the 25th and 26th November 1963.[3] The first part contains the following:

> At the moment a wide field is open to the productive development of psychology welded to the physiology of the brain and based on the results of social, natural, and technical sciences, in particular those of Cybernetics. [...] Psychology has reached such a level of development that not only is it possible to analyze the constituent elements of thought, but also to shape some of its aspects. [...] The rapid development of science and its irruption into the "holy of holies" of nature gives to the natural sciences a very clearly defined character of militant atheism. [...] The irrefutable scientific conclusions about the materiality of the world and the possibility of recognizing the inner capacity of matter for spontaneous movement and evolution penetrate ever more deeply into human awareness.

In contrast, we have the explicit spiritualistic declarations of the founder of cybernetics himself, Prof. Norbert Wiener. It is well known that Wiener was a spiritualist. However, in addition to what Wiener himself said about himself to me, here I merely mention the commemoration of Wiener given by Prof. Charles Arthur Muses of Lausanne in the periodical of our CNR "*La Ricerca Scientifica*".[4] In it he states "What Wiener deplored is precisely the divorce of technology and cybernetics from all ethical and morally responsible considerations".[5]

---

[3] Published also in the *Kommunist* 1964, n.1, pp. 23–46, with the title "Formation of a conception of the materialist world and atheistic education"; Italian version in *Aggiornamenti Sociali*, p. S. Fedele 4, Milan 1964.

[4] "*La Ricerca Scientifica*" vol. 5 n. 1–3, January–March 1965, pp. 3–9

[5] In this note I reproduce the conclusion of this article: "What Wiener deplored is precisely the divorce of technology and cybernetics from all ethical and morally responsible considerations. In his writings he gives repeated warnings about this divorce".

It is also interesting to keep in mind that the subtitle of the original English edition of the *Introduction to Cybernetics* (1950) is "Cybernetics and Society". In this work Prof. Wiener, although he diligently seeks the most optimistic conclusions in accordance with the facts, is forced to conclude at the end of the last chapter that scientific technology as it is developing at the moment is incompatible with a positive *ethos*. The serious consequences of such an incompatibility are clear. Cybernetics without ethics is illusion.

There is a third fundamental aspect of the spirit of Norbert Wiener. This third aspect directly concerns the very nature of man and the logical error of reducing man to a mechanism, so as to make our equations better, or worse, at making the automatic control of one human being easier for another. The definition of man as a mechanism, constituting as it does in itself an erroneous diminution through its very falseness, has nothing to say about ethics and thus leaves the vital question "Who shall control, and who shall be controlled?" to the law of the jungle. The idea of man as a mechanism thus leaves man with the same level of moral development that he had at the dawn of history and reduces his technological "progress" to an ironic tragedy.

Instead of a club, man now uses nuclear weapons, but all the basic concepts remain at the same stage of development, which is sadly primitive. His anxiety about the error that reduces man to a mechanism is probably the key to Norbert Wiener's whole way of thinking. The power and pertinence of this anxiety cannot be more clearly shown than by citing the erroneous thought of certain people who have omitted what Wiener found so necessary and what is confirmed so profoundly by logic and the facts. No quotation could better illustrate the impasse in which this school of thought finds itself than the following words of Prof. W. Ross Ashby of the University of Illinois (p. 66 of "Introduction to cybernetic systems" in the book *Induction: some current issues,* Wesleyan University Press, Conn., U.S.A.), which has recently come to my attention: "We

There is an equally religious tone in a widely read article by Vannevar Bush, who at the moment is honorary President of MIT at Boston Mass. in the U.S.A. This article[6] could very well be one of those read at the conferences of this cycle because it sets out precisely the future prospects that science has opened up to humanity. It presents the boldest predictions in a religious context, and I will read you a most interesting observation from it:

> The new materialism is in no way limited to those who propagate a communistic organization of society. The philosophy of existentialism, which holds such a fascination for young people, takes many forms, and in each one of them there is an interest in what existentialism claims is science; this often leads to a rigid atheism.[7]

This holds true for our western, so to speak, cultural and scientific environment.

4. Since our question is framed in this way, we can answer it with a series of assertions.

Cybernetics can favour atheism by force of indirect attractions. The first of these is the total absorption of a scholar's attention in the examination of immediate causes. He tries to delineate the wonderful complexity of natural structures just as an intelligent worker, for example, might investigate a new machine he admires in a canning factory. He exhausts himself in noting down all the different pieces of

---

are unable to find at any point in the processes of induction, deduction, prediction and decision, an example in which the powers of a human being are essentially different from those of a constructed mechanism that is sufficiently complex".

And Ashby's definition of a mechanism is a closed system whose states are absolutely determined or predictable. The error is clear, and the example that Ashby cannot find is under everyone's nose: man designed the mechanism, but the mechanism he designed did not make man. This essential difference is profound and clear. The other errors arise from ignoring it. Prof. Wiener would have deplored, ("Emotional preferences for gadgets (mechanisms) instead of for human beings are particularly dangerous at the present moment" and "The living being is not a closed system" cf. (15), pp. 16–17), the fact that the treatment of the argument of cybernetics was pervaded by a way of thinking that was so immature and confused, to the detriment of young students whose formation would inevitably be distorted by contact with such errors of a pathological logic.

Nature does not know machines. All-natural systems are not closed ones, but open ones at constant states ("steady-state energy"). Mechanisms, on the contrary, i.e. systems made by man, are closed systems. The so-called "mechanical" aspects of nature indicate an error of anthropomorphic description, because the pure mechanism does not exist in nature. Such aspects, after a sufficiently profound investigation, are properly only of biology in the strictest sense for the presence of living beings with the freedom of choice that this term implies.

Reciprocally, only man has invented the machine, and that which has been made can never, neither logically nor ontologically, contain the maker. Prof. Norbert Wiener, and this is his great merit, has never lost sight of this fact and this elementary and profound conclusion. To say this of him is what he himself perhaps would have considered the greatest tribute that can be given to a great man. Wiener has deserved and does deserve such a tribute. On Wiener himself see "*God and Golem, Inc. – A Comment on Certain Points where Cybernetics impinges on Religion*". The M.I.T. Press 1964.

[6] "Science Pauses" in *Fortune* May 1966, pp. 116–172, Italian version in *Mondo Orientale* n. 105, September–October, 1965.

[7] On page 40 of the Italian version.

the mechanism and goes away without giving a thought to the design office where it was invented, or to the businessman who ordered it, or to the director who keeps it going.

Indirect attraction is also that impoverishment of ideas that we call 'mechanicalism', which opens the door to materialism. It is a dangerous slippery slope, especially for those who do not have the chance to specialize deeply and who therefore regulate their science not on the basis of how much research they have done, but simply on the basis of how much they have read. Take for example a recent publication on the philosophy of automatons: you will probably have difficulty in finding any explicit assertions of materialism there, but you would notice that on many pages there is something that I cannot define better than the slippery slope that I mentioned just now. Perhaps the expert who writes is far from falling down it, but the incautious reader is in danger of doing so quickly.[8]

When cybernetics professes itself materialist, it does so through reflexion of mental positions that are prejudiced and outside cybernetics, i.e. not by virtue of data provided by cybernetics itself, but by force of presuppositions derived from mental philosophical positions prejudiced in another area. If you have read the article by Alfredo Todisco in the *Corriere della Sera* of 17th November 1965, you will have seen the following:

> Cybernetics starts with the assumption that a human being is a machine (one formed, moreover, piece by piece in the course of evolution with materials from earth), a machine that functions thanks to the mechanism of its organs and without the help of any immaterial metaphysical principle, which for that very reason cannot be quantified.

If that means merely that cybernetics is not a deductive philosophy but an experimental science, we would all agree. If, however, it really means that cybernetics starts with the assumption that a human being is a machine, we should absolutely say no. It is not at all true that cybernetics starts with the assumption that a human being is a machine: if it is a science, it starts only with factual data, and takes a human being for what that human being is. And the fact that the incredibly organized structure of that operative unit that we call a human being is formed by evolution is neither a starting point nor one of the recognized conclusions, nor is it one of the arguments or proper items of cybernetics, but it is its insertion from outside into a philosophical concept that has been prejudiced in another place thanks to the decisive factors of the *de facto* existence of the world.

Therefore, it is true that any treatment of cybernetics will be materialistic if its author inserts it into a materialistic context. You will tell me that it is also true that it will be spiritualist if it is set in a spiritualist interpretation of life. In other words: let us leave assumptions of various tendencies that are outside cybernetics and let

---

8 "*La Filosofia degli Automi*" ed. Vittorio Somenzi, Boringhieri 1965, is a small anthology of writings on automatic calculation, automatons, cybernetics. As a sample of what I allude to here, I offer the following quotation from Von Neumann p. 245: "This fact that the complication, just like the rest of the organization, is degenerative under a certain minimum level and above it can become self-supportive and indeed increasing, will probably play an important role in any future theory of matter".

us concentrate on identifying those supports that cybernetics offers from within its own specific contents.

However, I have not been able to find in publications about cybernetics any "demonstration" of materialism. There are only allusions and nudges, all of the sort mentioned before, i.e. through their setting in mental positions that are outside cybernetics. Obviously, if there were any, they would all be of a negative type, such as those of Michel Verret in a recent communist book[9]: i.e. by force of imaginable substitutes for God such as spontaneous motion, auto-evolution, causality, eternity and the infinity of matter. All are attributes of the divine and are as much beyond our immediate experience as are the attributes of God. We can challenge any atheist to escape from the following alternative: either there is a God creator, or the world and all of us are God. But if the atheist shifts within this alternative, he offends against the very logic of his own atheism. If he affirms that only the object of our experience exists, he should realize that absoluteness, eternity and infinity are not objects of measurable experience, but on the contrary, no more nor no less than that mysterious "beyond", recognition of which is the starting-point for religious meditation.

Cybernetics does not offer positive proofs of atheism or of materialism. (I link these two terms, not because they have precisely the same meaning—they do not—but because they are, in fact, always related). The fact that there is electro-biological support of the activity of thought is not a proof of materialism, because the binomial "body and soul" is not a contradiction, as if, since the body exists, the soul therefore does not exist and cannot even be imagined. But the body, by showing that it is moved by a force capable of organizing its activity, opens up the prospect to the following comparison: the world of the body is nothing if not an expression, diffused and diluted, of a world of intelligence that is the prime and necessary reality, and this reveals itself in the body just as an artist or an inventor does in their works.

I remember that in this place—to think that it was 19 years ago!—Prof. Nicola Pende gave a paper on the subject "Body and Soul" during the ninth course of Christology of this *Studium Christi.*[10] In Christian thought the binomial "body and soul" extends to that high point that is the dogma of the hypostatic union of the one nature of Man—soul and matter, and physical, chemical, biological, psychic and spiritual energy—with the person of the Word of God. The fact that any concept can be expressed in numbers is not a proof of materialism: as well as referring to the very learned exposition of Prof. Rivetti that we have just heard, I would also allude to the fact that no page of human philosophy has reached the splendour of that Trinitarian theology which finds the number within the unity and unicity itself of the only God.

The seeming resemblance between thought and any other activity is not a proof of materialism. Bodily activities are nothing if not "participations" or degrading

[9] *Les marxistes et la religion*. The title of the Italian version is "*L'ateismo contemporaneo*". 1963.

[10] See "*Il Cristianesimo e le Scienze*", Rome, Studium Christi, 1947, pp. 145–164.

diminutions, step by step, of the prerogatives of the spirit and of thought: just as in an orchestra the vibration of the string of an instrument is nothing if not a capillarity in which the spiritual power of a harmony that mysteriously exploded within the spirit and creative imagination of the composer is expressed and, as it were, channelled.[11]

5. Cybernetics, however, presents signs of the presence of the spirit and of the presence of God, although, let this be clear, cybernetics is not religious philosophy, i.e. it is not its specific task to occupy itself with them as the object of its research. To sum up, the signs of the spirituality of thought and of the soul appear as discrimination and difference between the associative facts from which derive conditioned reflexes and the capacity for organization as properly defined, or active programming as properly defined. The transmission of commands always starts from a centre and goes to a many-sided periphery.

When a human being devises an organization, we find its starting-point in this conception, i.e. in the sense of mental conception of an operating formula, which is like the embryonic cell of that organization. If you wanted to bring such a conception within the scheme of the associative facts, it would split the scheme. Such a mental creation is not a conditioned reflex but a force of thought, one and unitary, which subsumes within itself and dominates a multiplicity of things and events. This is what we call spiritual operation: a unitary force of multifarious control. Here cybernetics will meet the sign of the spirit whenever it is a question of deliberate and conscious operations. That is, when the commands are transmitted from the network of the nervous system to the rest of the body it will go back to the program loaded in this, our wonderful calculator, and from the program it will record the day and the hour of its conception.

The same picture can be seen when, instead of there being deliberation about an operating program, the phenomenon of interpretation is examined in the mind. Understanding a language, catching in a picture or in music the formal values of harmony and beauty, this faculty of seeing in one single corporeal fact, encapsulated in the four dimensions, the transparency of a beauty that is always itself for as many times as it can be repeated, this phenomenon too, which in our thought is its heart, heat, colour, and joy, is a unitary act of force of multifarious control, i.e.

---

[11] "*Dante nel Convivio*", (ed. G. Busnelli, G. Vandelli, Florence 1934 tr. III, cap. VII, par. 6, p. 335) commenting on verse 37 of the second canzone (p. 254) "*in lei discende la virtu' divina*" confirms that it was one of the usual concepts of the cultural environment of his time: "*E pero' che nell'ordine intellettuale de l'universo si sale e discende per gradi quasi continui da la infima forma a l'altissima e da l'altissima a la infima, si'come vedemo ne l'ordine sensibile; e tra l'angelica natura che è cosa intellettuale , e l'anima umana non sia grado alcuno, ma sia quest'ultimo a l'altro continuo per li ordini de li gradi, e tra l'anima umana e l'anima piu'perfetta de li bruti animali ancor mezzo alcuno non sia; e noi veggiamo molti uomini tanto vili e di sì bassa condizione, che quasi non pare essere altro che bestia; e così è da porre e da credere fermamente, che sia alcuno tanto nobile e di si alta condizione che quasi non sia altro che angelo: altrimenti non si continuerebbe l'umana spezie da ogni parte, che essere non può. E questi cotali chiama Aristotile, nel settimo del'Etica: e cotale dico io che è questa donna, sì che la divina virtute, a guise che discende ne l'angelo, discende in lei*"

catching formal unities as they are, which remain themselves and one, and enchanting, even though repeated and, as it were, reverberated in so many photograms in the running of the film of becoming physical.

6. Cybernetics meets the signs of the presence of God when it meets in man, not just and in the first instance, the operating programs formulated by his conscious deliberations, but, on the contrary, when it meets all those that are heaped up behind them. That is, those that are his nature and make him capable of programming responsibly as part of his own actions. Unfortunately, I am compelled by the time limit to give a summary, which I fear will be somewhat rough, and which will unfold in three ascending stages.

First. Examination of the organized complexity that is a human being provokes the intellectual "feeling" that behind this "marvel" of nature there is an incredible intelligence and force. Let us just look at the nervous system of a human being as if it were a machine; all right, let us just say that it is a machine; but (thank heaven!) let us also say that it is everything that it is and that it is as it is! A human being is an apparatus of—so we are told—one hundred thousand billion cells. Who put them together? According to what plans and patents? The nervous cells of the encephalic cortex alone are ten billion, all formed during the nine months of gestation: then their number does not increase, and starting from a certain age, alas, we start to lose them at a rate, if I remember rightly, of twenty or thirty thousand a day. They are connected among themselves by many filaments that can reach up to a metre in length and that tie themselves up into knots in just as many synapses.

All right, let us call it just a machine or a computer or a command centre: nature, then, is a factory of these machines, and you see how many she puts into circulation every day - so many that many people try to stop the excessive production. Well then, if our little factories always have an office, or at least a workshop of ideas, and then an entrepreneur and a director, on what basis should we be able to say that the world is without one? By what chance do spontaneous movement and auto-evolution mean disaster in our better factory, while, on the contrary, in the factory that is the world there should exist intelligent organizing powers?

Second. It is precisely by examining, above all, the logical content of this instinctive picture that reflection starts to be reasoned recognition—i.e. philosophy—of the presence of God. Or rather, that confused feeling that there must be something else at the roots of the world is first of all translated into a scientific analysis of the implications of organization in nature, which has its own laws, or programs, of evolution. How often have I felt excitement whilst reading the often-unconscious philosophy that is manifested in the publications of the experts of these techniques!

The "exits" towards the recognition of the presence of God are remarkable and impressive: and precisely because information theory, science of government, and cybernetics are essentially nothing if not the analysis of the phenomenon of active organization, examined in its downward progress when it should be the other way around. That is, from the result towards the first dynamic principal, how is it not possible to understand immediately that all the complex periphery nonetheless always has a centre, and one only, which is its motive force, and to be its motive force can it not also be its inventor? That is, if there is organization, there is a program, and if there is a program, there is a programmer? And if this programmer exists, it is the unitary principle of the power of the multiple? Well then, this is what we mean when we talk of "Spirit".

Third. With that, philosophy has already become natural theology, in the sense of the ability to read God in nature, and by nature, I mean both the experimental corporeal world and our own conscience. But it is full philosophy and mature natural theology when it goes on to notice the reality of organization in the universal woven fabric of being itself, i.e. in the structure or composition that lies within every known reality. Everything that we know is the realization of an operative formula: existence—essence—operation!

When the old scholastic philosophy repeated "*omne compositum habet causam*" (every composite thing has a cause), it said something that we should translate as follows: every organized reality has or is a program of a programmer. Every expression, every use of significative symbols is a reality that is organized and programmed and *vice versa.* In this respect, philosophy of existence flourished in rational analysis of the logical premises of inner life, mental ascent and mystical exaltation in accordance with which man first became aware of having God as a guest in the deepest roots of his being, then to understand—or, perhaps better, only to have a vague idea—that he was the host of eternity, and to find himself, an agitated, anxious, hungry, suffering, hopeful, disappointed being, within the heart of God.

7. Cybernetics, therefore, just like any other science, will push human beings of tomorrow against two opposite attitudes with regard to religious thought. In what proportion? It is as well to underline a feature of our present-day culture: the eclipse of the sacred. This is a phrase that I have taken from an interesting recent volume of religious sociology,[12] and I translate it, as far as it is of interest to us, with the words: laicity of science. The knowledges that are useful, or are motors or principles of programmed action are those that have become habitual. Those ideas that are frequently repeated by the means of communication become habitual. Unfortunately, the place of science in thinking about God, i.e. making clear the "exits" towards God that occur in all research on reality, is no longer frequently acknowledged to-day, yet it was taken for granted in the days of Dante and even later in those of Galileo and Newton.

Galileo in a letter of the 15th January 1633[13] to Elia Diodati emphasized this idea in an extremely polemical context: "*Il mondo dunque son le opere, e la Scrittura son le parole, del medesimo Dio*" (thus the world is the works, and Scripture the words of the same God).And in a letter of 1615[14] to Madame Cristina of Lor-

[12] SABINO S. ACQUAVIVA: *L'eclissi del sacro nella civilta industriale*, Edizioni di Communita, 1961. See also *Christians in a Technological Era*, Seabury Press, New York 1964.

[13] Ed. naz. Vol. XV, Florence 1904, p. 24.

[14] Ed. naz. Vol. IV, Florence 1895, p. 239.

raine he wrote "*La Gloria e la grandezza del Sommo Iddio mirabilmente si scorge in tutte le sue fatture e divinamente nell'aperto libro del cielo*" (the marvel of the glory and the grandeur of Almighty God can be seen in all his works and divinely in the open book of the sky).[15]

Unfortunately, we are no longer taught to read things as words, solid and four dimensional if you like, of God. Prof. Polvani referred to St Paul's saying that each thing has its own voice. But today we are all affected by a malformation: when we enter the room where nature expands its symphony, we bring with us our instruments to record its vibrations from a mathematical and physical point of view, and we go out with a trace of them on a piece of paper, but without taking them to heart for their beauty! It is this and nothing else that makes us slip into that practical atheism which is often called—just think of the paradoxical harshness of this expression—scientific atheism. It limits us to measuring what moves under the circle of the microscope's light and does not allow us to surmount the borders that form the boundary with mysterious darkness. And yet, that narrow circle of light is nothing if not a stage whereon we ourselves are placed to act our part, we who have come out from the wings of the mystery and into which we shall soon be called to return.

---

[15] Also, in the third letter to Signor Marco Velseri about sun spots: "*a poter meglio filosofare intorno ad altre piu' controverse condizioni delle sostanze naturali; le quali poi finalmente sollevandoci all'ultimo scopo delle nostre fatiche, cioè all'amore del divino Artefice, ci conservino la speranza di poter apprendere in lui, fonte di luce e di verita', ogn'altro vero*".

And in the *Dialogo* about the two greatest systems of the world (ed. naz. Vol. VII, Florence, 1897, p. 130) (these are the last words of Salviati of the Giornata Prima): "*Concludo pertanto, l'intender nostro e quanto al modo e quanto alla moltitudine delle cose intese, esser d'infinito intervallo superato dal divino, ma non però l'avvilisco tanto, ch'io lo reputi assolutamente nullo, anzi quando io vo considerando quante e quanto meravigliose cose hanno intese investigate ed operate gli uomini, pur troppo chiaramente conosco io ed intendo, esser la mente umana opera di Dio, e delle più eccellenti*".

For Isaac Newton see *Optices Libri Tres*, Latin version, Padua 1773, in L. III, qu. XXXI and last, starting especially from p. 164. On p. 166 he concludes: "*Nam quatenus ex philosophia naturali intellegere possimus, quaenam sit prima rerum causa et quam potestatem et ius Ille in nos habeat et quae beneficia Ei accepta sint referenda, eatenus officium nostrum erga Eum, aeque ac erga nosmetipsos invicem, per lumen naturae innotescet*".

See also E. Pemberton: *Saggio della filosofia del Signor Cav. Isacco Newton*, Italian version, Padua 1733, for example the conclusion pp. 215 *et seq.*, where he summarizes the religious thinking expressed by Newton at the end of his *Philosophiae Naturalis Principia Mathematica.*

It is the specific responsibility of us ecclesiastical scholars to ask ourselves what can be done in order that human beings of the future, in the common equipment of ideas familiar to them, may have once again have, as scientific culture had until a couple of centuries ago, a sense of the sacred.

Chapter 9 Experienced-based Results with Preparations for the Use of Automatic Calculation in Biology

Roberto Busa S.J.

First published as: Busa, R. 1966. Risultati ed esperienze di fatto sui preparativi previi all'impiego del calcolo automatico in biologia. In *Atti del Convegno Internaz. sulla Sclerosi Multipla, Gallarate 24–25 Aprile 1966*, ed. R. Montanini, 377–382. Gallarate.

Fig. 9.1. Busa pictured at the International conference on Multiple Sclerosis, which was held in Gallarate, 24–25 April 1966, where he presented the paper below (Busa Archive #0590).

Editors' note

This text was delivered at the International Conference on Multiple Sclerosis in Gallarate in 1966. In it Busa seeks to build common methodological ground with disciplines like biology, which are distinct from his but similar in their desire to use computing to assist in the advancement of knowledge. He also aims to problematize popular portrayals of computers as "giant brains" that make tasks easier and quicker to complete. He writes again that they are in fact "high speed cretins", of the great deal of intervention that is required in order to use them effectively and of the intense frustration that he has felt in the course of utilizing them.

Regarding methodology, Busa writes that irrespective of discipline, every use of computing in research must be built on the following stages: analysis of source material (also referred to as pre-editing in his other publications), grouping, codification, input and

J. Nyhan and M. Passarotti (eds.), *One Origin of Digital Humanities*,
https://doi.org/10.1007/978-3-030-18313-4_8

verification. In discussing the continuum from codification to standardization he writes how the desire for standardization of linguistic codes had already been expressed by 1960.

Busa argues that the use of computers in research has resulted in fundamental changes to the ways that research is designed and undertaken. This is reflected in what he sees as a "numeric turn" (or perhaps a symbolic turn) which he describes as the "extension of mathematical calculation and formulation to all areas". He believes this to be emblematic of his times, as shown by the take up of such approaches in unexpected contexts, for example, pastoral theology. He discusses some of the other changes that computing has heralded by drawing attention to the new kinds of posts that it requires. He describes the post of a so-called "connection officer", or a person who can adequately explain their subject area to a programmer (he describes the process as being unidirectional from officer to programmer). Busa writes that computing prompts such profound change that it impacts even those who work at the periphery of an organization.

Your amazement at seeing me as a speaker at a congress of such highly specialized medicine will certainly be no less than mine! The reason for my being here is that I was invited to make public the advice I gave to the scholars of Gallarate when they told me that they intended to make a statistical survey of the sector of multiple sclerosis in order to run automatic calculations. My experience is in a completely different sector. It is in linguistics, in which we have worked on nine different languages in four different alphabets: Latin, Greek, Hebrew, and Cyrillic. Yet, it is obvious that such experience has commonalities with any other experience of the use of computers. Therefore, I should like to illustrate what preparation is necessary and sufficient for the use of automatic calculation: obviously that which is common to the use of computers in any field, from linguistics to physics, and from psychology to medicine.

1. The first thing that I told them was that they were quite right to use a computer, for this has become an indispensable instrument for scientific research. Young scholars must start to use this for conducting their research scientifically, and it does, indeed, introduce a new dimension. There are two parts to this: first there is the use of the computer, and then there is the increase in the use of statistical investigation and mathematical calculation. These latter reach into domains that were previously not thought of. You know that today even pastoral theology, or the science of the cure of souls by us priests, is expressed in algebraic formulae as one of the areas of religious sociology! I repeat, one of the specific characteristics of the scientific research of our time is precisely the extension of mathematical calculation and formulation to all areas.

This fact has perhaps attracted less public attention because of the irresistible appeal made by the marvellous achievements of automation technology, mainly thanks to electricity, but it is not for this reason that it is less remarkable. This phenomenon is actually still expanding: the older generation regards it partly with amazement and dismay. But an ever-increasing number of young people realize that the use of the computer is becoming just as indispensable in a centre of research as an automatic washing machine is in a home.

2. The next thing that I told them was that, if anyone should chance to say that the computer threatens to rot the human brain, they should send him to me! And I would tell him that it was quite obvious that he had never worked with computers! Computers extend into the future the promise made by the Lord in the book of Genesis when he said that by the sweat of our face we would eat bread (III 19). They multiply the urgency for intelligent work by man: they condense a greater quantity of organizing and interpretative work into a smaller space of time.

Using a computer simply means working more intensely than before: I am sorry to have to say it, but it is so!

3. Thirdly. For some reason the computer has been called a giant brain. I am most annoyed with whoever invented and publicized this phrase! I will continue to say: would that it was indeed a brain! Unfortunately, the computer needs to be programmed, and here I come to the main part of my argument. The computer is nothing more than a combinatorial machine: the French have a very clever word for it "*ordinateur*". A German professor once described the computer to me as "*Hochgeschwindigkeittrottel*". Literally translated this means "high speed cretin"! It is a machine that plays a great game of solitaire with units of punched cards or units of records, but only in accordance with the programs that have been prepared for it. Depending on the particular cases, the programming of a computer requires from several hundred to several thousand man-hours.

4. What does programming mean? Let us divide it into two: the specific programming which is the technique of the formulation of the circuits and operative cycles of the computer, and the general architecture of processing, which is sometimes called analysis. The analysis, in turn, can be divided into two, for it is the collaboration of two competences: that of electronic calculation, and that of the specialist in the discipline in which the research is to be carried out. In other words, I stress that in any research carried out on a computer there must be, as it were, a connection officer: someone who both knows the subject and who can also translate its technical terms into a language that can be used by the programmer.

5. The analysis is carried out in three stages. The first is analysis proper, i.e. putting the material into elementary units of information; this is because the computer is nothing more than a combinatorial machine. To put data into elementary information units is an exquisitely philosophical work of classification. To the concept of an elementary unit, as defined by Aristotle, ("an element is something that cannot be subdivided into parts of different sorts"—5 (4) *Metaphysics*, 1014a 26) must be added the level within which it is to be contained. This is for our peace of mind: to avoid being submerged in a series of problems that plumb the very depths of metaphysics. For example, in linguistics at certain levels of research the elementary unit is the sentence, at others the syntagma, at others the word, and yet at others the morpheme or the grapheme. Naturally the unit must always be categorical, i.e. proper to the category of information it deals with.

6. In the second stage these units of information are put into groups which, while they are compossible with each other, still contain within them units in alternation.

In fact, both the punch card and the magnetic tape can be subdivided into "zones". In each zone the data contained within it must be in alternation, so that of all the possible data only one comes up. If, however, one is dealing with pieces of information that can all be true together at the same time, these cannot be put into the same zone, but must be distributed throughout different zones. For example, sex and age cannot be put into the same zone, for they are not in alternation, but compossible. In practice, this is one of the most delicate points at the beginning of an automatic calculation. I repeat: to put elementary units into groups that are compossible with each other, yet each of which contains information in alternation.

7. The third stage is codification. The main problems with codification are those of operation. At the level of the speciality with which it is mainly concerned, the problems occur only with regard to standardization: it is both ideal and desirable, yet difficult to put into practice.

8. Standardization means that all those who carry out statistical research, say, for example, on the subject of your conference, must adopt both the same numerical or alphabetical codes for the same units of information, and the same zones. In the conference that I organized at Tübingen in 1960 I was asked to standardize the codes for the automation of linguistic analysis for the whole world. My reply was: "How can I succeed in this, when I know that the Lord, my Master above, tried to standardize something that interested Him much more, namely religion, and see what success *He* had?!" I would say that standardization is a problem that transcends human capabilities. However, the problems of codification, if they are not to do with the standardization of a scientific discipline, occur only at an operative level and may reasonably be entrusted to the programmer, as properly defined.

9. After the analysis of the problem comes the phase of installation, which we call perforation or the preparation of input, i.e. the computer's input material. Here the difficulties are simply practical ones; they do not involve any abstruse theory, but they arise from a pitiless practical law: the law of increasing entropy, the irrepressible possibility of human error. In our linguistic research the times of input production are as follows: one quarter, 25%, of the time is for the initial production, the other three quarters, 75%, is for checking and correcting. Even so, not a week goes by but we find mistakes in pages that we have carefully read ten times over!

10. This is the main reason why I recommended my friends in the Centre for Multiple Sclerosis at Gallarate to attach great importance to something which in itself is banal, *viz.* the type of component on which the research data are to be recorded. I suggested to them that this component should be arranged typographically in such a way that it could be used as a perforation document without any further transcription, for it is from this that the secretary will punch the data onto the cards, which will then be loaded onto the tape. I also warned them to multiply by four whatever time and money they had estimated for the transcription: 25% for the first transcription, and 75% for tracing and correcting the errors made during it.

11. The use of the computer has another important concomitant. It necessitates the modification of methods and speeds by every one of those who are involved in any part of any organization. When a computer enters an organization, it compels even those who work on the periphery to go at its pace. To be exact, this means an acceleration, a condensation, and even a qualitative increase of methods and rhythms of work: in other words, it involves a number of difficulties that were previously unsuspected by those whose knowledge of the computer was confined to what they had read in the newspapers. It is not at all true that it is enough merely to press a button. The rocket does indeed take off when the button is pressed; but this is the result of months and perhaps years of work by many people, work that is accurate in the minutest detail.

12. With these two points—the necessity of analysis and the necessity of working out carefully how to deal with the installation—I have come to the end of the advice I thought worth giving. Yet I would add four brief afterthoughts.

Another methodical rule essential to these works is that of beginning, and then that of continuing, and then, if your experience is anything like mine, that of knowing that your work will eventually turn out well after you have been sitting half a dozen times on the stairs of your laboratory, with your cheek laid pensively on the palm of your hand, feeling an urge to commit either dismal suicide or mass murder!

Lastly, so as not to burden your soul with too heavy a weight, I would suggest that you prepare a phrase-book of innocent but effective exclamations, graded like the Mercalli scale for measuring earthquakes, to refer to at critical moments in your work!

Chapter 10 The Function and Use of an Electronic Computer

Roberto Busa S.J.

First published as: Busa, R. 1967. Funzionamento e uso di un calcolatore elettronico. In *L'uomo e la macchina. Atti del XXI Congr. Naz. di Filosofia. Pisa 22–25 Aprile 1967*, vol. III, 240–245. Torino: Edizioni di Filosofia.

Editors' note

In this article Busa explains what an electronic computer is and what it does. He also returns to the now familiar theme of the limitations of computing and computers. He writes that the computer is composed of three elements: input, central processing unit and output. Regarding input, for example, he describes how the input material can take the form of cards, magnetic tape or punched paper tape. On cards, letters are indicated by specific combinations of holes made in the columns of the cards; on the magnetic tape, which is plastic tape coated in iron oxide, the perforations are magnetized. The read/write heads that the tape passes under can then transmit an electric current when they detect a magnetized zone.

The tapes available to Busa when he wrote this could store a maximum of 22 million characters; they could be run at a speed of 10,000 words per second. Nevertheless, he emphasizes that when one works with a computer one is merely directing electrical impulses to specified ends and this is nothing compared to the wonder of the macrocosm.

In earlier articles we have seen Busa indicate that the disambiguation of homographs was done by hand. Here he refers to the machine dictionary that had been devised to assist in this process and we see that it has hastened it dramatically. The work on the machine dictionary involved codifying by hand some 150,000 Latin words and annotating them with an average of 40 symbols, followed by the necessary steps of keypunching, checking and correcting the material. Later, Busa how "a team of ten priests worked with me for two full years to design a Latin machine dictionary. Called the *Lexicon Electronicum Latinum* (LEL), it is a set of tables by which a computer is able to lemmatize the words of Latin texts" (Busa 1980, 86).

References

Busa, R. 1980. The Annals of Humanities Computing: The Index Thomisticus. *Computers and the Humanities* 14(2): 83–90.

First of all, let me assure you that I will spare you a repetition of what I wrote in the fascicle published for this occasion by CNUCE entitled "Electronic Elaboration in Linguistic Analysis".[1] In fact, I thought I would do you a favour by explaining in everyday terms what an electronic computer is and how it works, that is, by telling you what someone who does not understand it has understood about

[1] [Editors' note] The fascicle referred to here and elsewhere in this text is L'*Index Thomisticus* della Facoltà di Filosofia dell'Aloisianum di Gallarate, in *L'elaborazione elettronica nell'analisi linguistica*, pp. 1–23, Pisa, Centro Naz. Univ. Calc. Elettron., 1967.

J. Nyhan and M. Passarotti (eds.), *One Origin of Digital Humanities*,
https://doi.org/10.1007/978-3-030-18313-4_9

it! We can split the electronic computer up into three parts: input, central processing unit (CPU) and output.

I Input unit: the computer reads

Let us speak about the unit of input of the computer. The material that arrives at the computer, on which and with which the computer works, is either cards or a magnetic tape or, for certain types of computer, a punched paper tape. The spools of magnetic tape that you have seen in the room of CNUCE[2] are precisely the input of the computer. To get to the magnetic tape one generally, but not always, starts from the punched cards. In [Figure 10.1] of the fascicle you will find a precise description of the transposition of the alphabet printed in a book to the alphabet punched onto a card and thence to the alphabet magnetized on the magnetic tape.[3]

For example, the *Corpus Thomisticum* of St Thomas Aquinas contains about one million seven hundred thousand lines. This number of lines has been copied, letter by letter, together with punctuation and spacing, with the addition of some other small details, especially on punched cards. I should add that we had to recopy it practically six times: twice in order to write it out, once in order to punch it, and thrice to get rid of the errors made in punching. When the whole of the text, just as it is found in the editions, had been put onto cards, these cards were inserted into the computer, which then copied them onto a magnetic tape.

The alphabet of the cards is composed of holes, which have meaning according to their vertical combination, which we call a column. For example, position 12 and position 1 in the same column are the letter A; position 12 and position 9 in the same column are the letter I, and so on. The holes of the cards on magnetic tape become kinds of magnetized notches, dashes, or rectangles. They are not impressions like those of sound recorders; they are discontinuous and like dashes. [Figure 10.1] is a facsimile. These dashes are not incisions, but little magnetized zones of magnetic tape, the magnetic tape being a plastic tape coated with iron oxide. Thus, we have the same text, the same words, the same content expressed with other symbols, and these are the symbols that the computer reads.

To say that the computer reads them, means that the magnetic tape passes under the read/write heads; these sense the presence or absence of a magnetized

---

2 [Editors' note] Here Busa is referring to the Linguistics Division of the Centro Nazionale Universitario di Calcolo Elettronico (CNUCE) in the University of Pisa. Antonio Zampolli, who had worked with Busa in CAAL in Gallarate, founded CNUCE in 1968 (see Zampolli, A. 1990. Summary of the activities of the Istituto di Linguistica Computazionale. *Computers and the Humanities* 24(5–6): 411–416). Busa moved his operations to Pisa after IBM had wound down his centre in Gallarate, in late 1967. Zampolli would, of course, also become a major figure in humanities computing and computational linguistics.

3 [Editors' note] Figure 10.1 is reproduced in the appendix that the editors of this volume have appended to this paper.

zone, and where they find magnetization, they transmit a very small electric current which amplifies it.

To give you some data: the greatest of the three possible densities on our computers is 800 characters per inch, circa 25 mm. In a spool of a magnetic tape of normal dimensions there are circa 900 metres of tape. At this greatest density the tape could, if there were no need for gaps, contain up to 22 million characters. The speed of the reading, i.e. the speed of running the tape, which is useful in enabling the computer to feel the characteristics of the individual magnetized notches, is that of 90,000 characters per second, which corresponds on average to 10,000 words per second.

Let us pause a while to consider what a marvel of technology these machines really are: yet another example of how in our days man has been put in complete control of what is infinitely small. The only thing that man can do with the computer is to direct mere electrical impulses to his own ends, and these are nothing when one thinks of the boundless space of the macrocosm. Yet the computer still needs to read what "the programs" are from punched cards. Allow me to explain.

II CPU; the computer counts and compares

Suppose we have the words of a text, one after another, on a magnetic tape. On the basis of the programs the computer separates the words from each other into so many units, which we call *records*, and which are the elementary units of elaboration. Secondly it arranges them in alphabetical order. How? The "instructions" of the program have arrived in the operative memories of the CPU. The program is a series of cards whose perforations successively command the closing and opening of the circuits of the CPU.[4] The program calls up the first and second words from the spool of the text [sic] to the CPU and compares them. The "comparison" is a fact of electronic physics: identical electric situations (strictly speaking "magnetization directions"[5]) exhibit in certain mechanisms a behaviour different from that exhibited by unequal situations. In this way, at the end the series of small magnets or charged elementary small rings which represents the word, for example *casa*, corresponds to a lower number, while that which represents the word *cosa* corresponds to a higher number. Thus, the instructions of the program open the circuits that cause first the word *casa* to be "written" on another tape, and then the word *cosa*. So, the computer, always on the basis of the instructions, calls up two successive words, compares them, arranges them in alphabetical order, and then writes them on the output tape. Then it takes another two and so on.

The time needed to call up these words in the "registers" of the CPU is called access speed. The 7090 has an access speed of 2.18 microseconds, or rather two

---

[4] [Editors' note] In the Italian text the term "*unità di calcolo*" is used here.

[5] [Editors' note] The term "*versi di magnetizzazione*" is used here.

millionths and eighteen hundredths of a millionth of a second, which means 500,000 words of 10 characters every second.

As you can see, the great strength of the computer is its speed. When it tells you that it is calculating exponents or square roots, it tells you with ... mental limitation! In fact, for the computer to carry out calculations of exponents is merely to perform a series of sums, but it does this so quickly that it appears to be one single operation. I recall that somebody in Germany once said to me: "Why do they call the computer a giant brain? I'd call it *Hochgeschwindigkeittrottel*!". If you will permit me, I will translate this literally as "high speed cretin"! But when more than 200,000 additions of 10 figure numbers can be done in one second, we may permit ourselves the luxury of saying that multiplication has been carried out!

In the CPU there is another type of memory, which is characteristic, and made up of networks of intersecting wires, and at each intersection a small ferrite ring is inserted. I will explain the little I understood. A small train of electrical impulses arrives at those wires, impulses shunted, for example, from the magnetic notches of the tape of the text. The impulses insert themselves into the ferrite rings and remain there, roosting like swallows on the light wires. Another little train of electrical impulses arrives, inserts itself and remains there. If, however, you shoot a third train from a certain direction, all that swarm of electrical impulses run off in a line, one after the other, and move along the course that you have already set up with the program. It is with this memory that the computer makes its comparisons, on materials derived from other "memories", which are the magnetic surfaces, in order then to transcribe again the results of its elaborations on other tapes.

Then, when in the linguistic elaboration, we have put the words of the text into alphabetical order (the 7090, if I remember rightly, can arrange a million per hour), we compare them with the so-called "machine dictionary". If ever the day comes that you take it into your heads to compose a machine dictionary, think of it as an excellent way of expiating your sins! We had to codify 150,000 Latin words by hand-writing on average 40 symbols for each. This equals 8,000,000 symbols written by hand, punched, checked, listed, corrected ... and still today we find mistakes in them! The machine dictionary is, first on cards and then on tape, a list of words, each one of which is codified with an appropriate series of codes. Of all the different ones that exist I mention only the so-called morphological codes, by which the word *faciebant* has the number code of the verb *facere* and other codes that say that it is the third person plural of the imperfect indicative active. Of course, we fell headlong into the *mare magnum* of homography and polymorphism! This machine dictionary has enabled us to codify by machine 10,500,000 words, having codified a mere 150,000 by hand. When the words are in alphabetical order, the computer compares them one by one with the words of the machine dictionary. If, for example, it finds a thousand examples of *faciebant*, to each of these it attaches the codes that the word *faciebant* has in the machine dictionary, always by processes of open and closed circuits, on the basis of the cards of the program.

III Units of output: the computer writes

The outputs, finally, are the way in which the computer expresses its own results. I will give you three examples. The computer can write on other tapes. Writing on tapes means that seven read/write heads, under which passes a tape, imprint on it as many magnetic notches as are ordered by the electrical impulses that come from the memories of the CPU. The speed of writing is like that of reading: 10,000 words per second.

Secondly, the computer can punch other cards. I will show you afterwards in the machines room which are the units that do the one and which the other.

Thirdly, the computer prints on rolls of paper with machines—*high speed printers*—that aim at the fastest possible printing speed. The last time I was in the USA IBM was undertaking the first trials of a printing machine that printed at a rate of 300,000 lines an hour: the difficulty that they were then trying to overcome was that as the paper unrolled at such a speed it became dangerously hot.

IV Conclusions: the computer accumulates human intelligence

To recapitulate: in the electronic computer we have units of input, a CPU and units of output.

In the units of input the "memories" are chiefly magnetic tapes: they contain and preserve the material on which the computer works. Besides this type of memory there is another that is part of the CPU. This is the operative memory, the operating table onto which the computer calls the various elements to isolate them and reassemble them. The CPU works on the basis of programs.

As I showed at n. 8 p. 3 of the fascicle of CNUCE, a computer without programs is like a chess-board without a player: i.e. a computer is merely a condenser or accumulator of human knowledge. It is, in fact, a systematic organization of operations, an organization which is, so to speak, put in a box in the form of programs prepared by man with the sweat of his brow. Every time I hear people speak of a thinking machine, I simply feel annoyed: it is not possible to make one! And if there were Saints who specialize in obtaining impossible graces, I would ask them to give me a machine which would create programs for me that had already been tested and that would work as soon as they were loaded into the computer.

I would also assign another task to a hypothetical, albeit impossible, thinking machine: precisely to track down for me chance human error. In automatic calculation we are obsessed with the possibility of error, for in linguistics those quadratures that guarantee the validity of numerical calculation are not possible. The computer has practically a negligible possibility of error. We human beings, however, have such a capacity for chance error; it cannot be suppressed, and is anything but negligible. In the fascicle of the CNUCE n. 42 p. 22 I made the following prophecy: when by God's good grace we shall have printed those eighty volumes of the *Index Thomisticus*, all the size of the volumes of the *Enciclopedia Italiana*, the very first time that any one of you opens any volume, at the very first glance you will find on the very first line a very grave error. Yet I must have run my eye over that same page seven or eight times without noticing it. I promise and assure you that this will happen!

The chief worry with the use of computers is not that of knowing how much they can do, but of knowing how much we can make them do; that is to say that here we must take care to have a method capable of foreseeing and controlling all the future behaviour of the machine, even that which is called chance or creative. Because of this need for method and coherence, I have often said that with the electronic computer, not only do we have a work tool but also a judge, or, if you like, a policeman, who demands absolute severity and honesty of scientific method, without exceptions and without approximations.

Now I will say no more, and allow Prof. Duro to speak. When you hear from him of the dimensions of the splendid project of the *Accademia della Crusca*, you will understand why I feel as though I were a tug-boat, a small squat boat that is mere force, which has just pulled a great ship out of port, and has the satisfaction of seeing it set off across the open sea. The *Index Thomisticus* deals with ten and a half million words. Prof. Duro will now tell you how many dozens of millions will be dealt with by the program of the *Accademia della Crusca*.

I thank you for your attention.

Appendix

First published as: Table II of Busa, R. 1967. L'*Index Thomisticus* della Facoltà di Filosofia dell'Aloisianum di Gallarate. In *L'elaborazione elettronica nell'analisi linguistica*, pp. 1–23, Pisa, Centro Naz. Univ. Calc. Elettron.

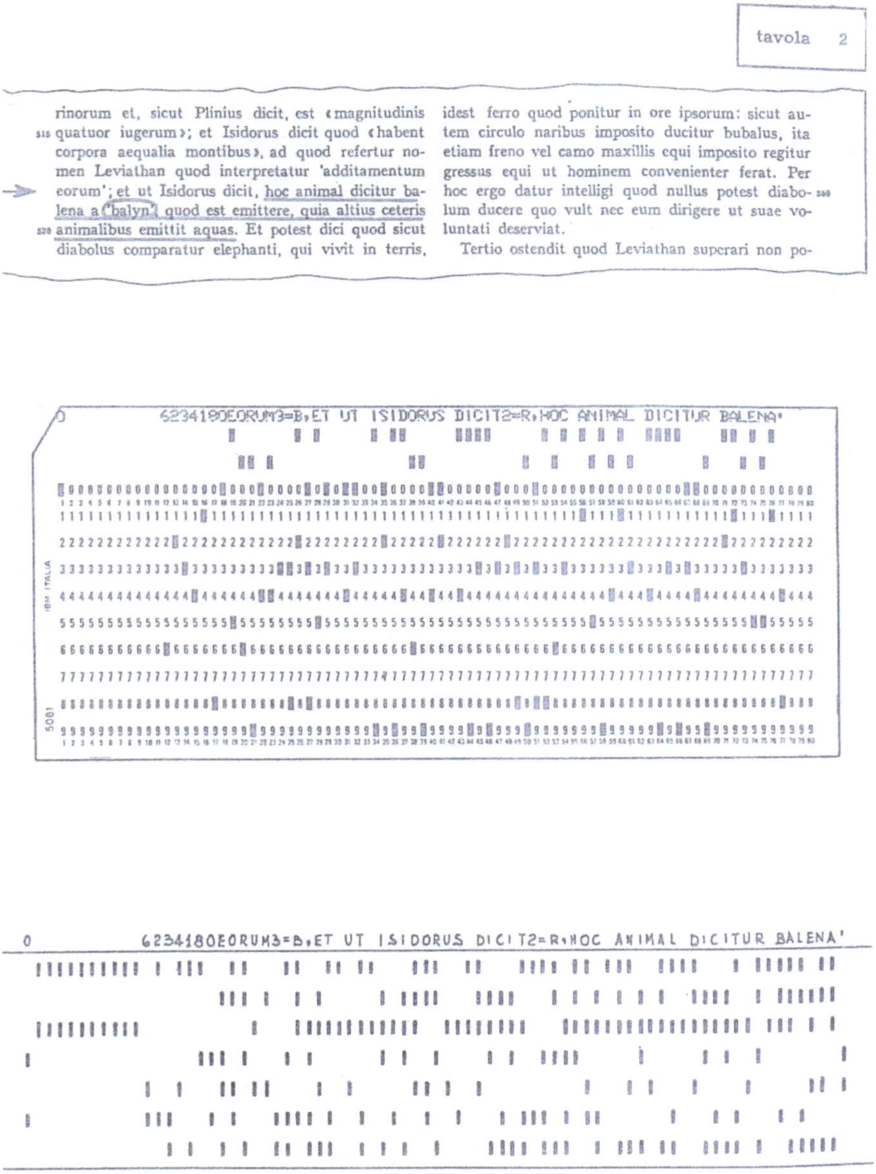
tavola 2

rinorum et, sicut Plinius dicit, est «magnitudinis quatuor iugerum»; et Isidorus dicit quod «habent corpora aequalia montibus», ad quod refertur nomen Leviathan quod interpretatur 'additamentum eorum'; et ut Isidorus dicit, hoc animal dicitur balena a 'balyn' quod est emittere, quia altius ceteris animalibus emittit aquas. Et potest dici quod sicut diabolus comparatur elephanti, qui vivit in terris, idest ferro quod ponitur in ore ipsorum: sicut autem circulo naribus imposito ducitur bubalus, ita etiam freno vel camo maxillis equi imposito regitur gressus equi ut hominem convenienter ferat. Per hoc ergo datur intelligi quod nullus potest diabolum ducere quo vult nec eum dirigere ut suae voluntati deserviat.

Tertio ostendit quod Leviathan superari non po-

Fig. 10.1. Transposition of the printed text onto a punched card and thence to magnetic tape

Chapter 11 Human Errors in the Preparation of Input for Computers

Roberto Busa S.J.

First published as: Busa, R. 1968. Erreurs Humaines dans la Préparation de l'Input pour Ordinateurs. In *Les Machines dans la Linguistique. Colloque Intern. sur la mécanisation et l'automation des recherches linguistiques*, 279–284. Prague: Acad. Tchécoslovaque des Sciences.

Fig. 11.1. This picture shows some of the young women who worked as keypunch operators in Busa's lab in Gallarate, 20/06/67 (Busa Archive #0613).

Editors' note

In 1976, Busa wrote how the *Index Thomisticus* was "a product of much work of a large team; for five years I had up to sixty-five co-workers. We estimate that it entailed one million man-hours" (Busa 1976, 1). To the best of our knowledge, the article below includes one of the few sustained discussions of the work of the female keypunch operators who were among those co-workers. They transposed onto punched cards the texts that would be processed by Busa and others in CAAL (see Terras and Nyhan 2016; Nyhan and Terras 2017).

Busa begins by indicating that a tremendous amount of data had been captured in the preceding years, some 12,000,000 million words in four alphabets. Though he does not necessarily make this claim, his description of the steps involved in punching the text implies that it must have been quite skilled work. Busa mentions how the keypunch operator's work required some level of interpretation of the category of textual feature

J. Nyhan and M. Passarotti (eds.), *One Origin of Digital Humanities*,
https://doi.org/10.1007/978-3-030-18313-4_10

they were transcribing and that they were required to know complex combinations of keys to encode special characters, and the like. That the keypunch operator's work was skilled rather than merely clerical is something that cannot be inferred from most of his other publications. Human error, Busa writes, could occur during any of the four stages of the human-led parts of the process: printing errors (presumably this refers to errors in the texts from which the project was excerpting), pre-editing, punching and lemmatization. Despite this, in the rest of the article he focuses almost entirely on mistakes that were made during the keypunching of the text.

Works Cited

Busa, R. 1976. Guest Editorial: Why can a computer do so little?, *ALLC Bulletin* 4(1): 1–3.

Nyhan, J. and M. Terras 2017. Uncovering 'Hidden' Contributions to the History of Digital Humanities: the Index Thomisticus' Female Keypunch Operators. Paper presented at Digital Humanities 2017, Montréal, Canada.

Terras, M. and J. Nyhan. 2016. Father Busa's Female Punch Card Operatives. In Matthew K Gold and Lauren Klein eds. *Debates in the Digital Humanities* 2016. Minneapolis; London: University of Minnesota Press. 60–5

1. We have already put onto punched cards, for electronic elaboration, natural texts containing 12,000,000 words in 9 different languages in the Latin, Hebrew, Greek, and Cyrillic alphabets, which deal with different, subjects, periods, and cultures, such as the Qumran manuscripts, the works of St Thomas Aquinas, and abstracts of nuclear physics. It is of the utmost importance for us to rid the entries of human errors. This is also the case with other centres that plan to automate the preparation of an index to establish an historical dictionary.

2. This work in linguistic elaboration answers immediate linguistic needs, but it seems that it will also be a valid contribution for those who are working in other directions, if it remains true that, for example, the automation of scientific information cannot be only a practical operation, because of the lack of knowledge of the factors and parameters of language. It is true that research in phonetics leads to solid results, even if it is performed on random samples. But for linguistics, I mean both structural and semantic, the analysis for a complete inventory and the putting of tens of millions of words of natural texts into an index seems to be absolutely necessary.

3. The necessity of avoiding and finding errors is a very pragmatic and perhaps plebeian task—I would even say a very humble task. But it is demanded by a pitiless law: human fallibility needs a diagnosis and a treatment in order to be cured, and an education in order to be avoided.

4. Amongst the many possible levels of communication techniques where human errors could occur, four points concern us in particular: the detection of printing mistakes, preliminary editing [Editors' note: also referred to as pre-editing elsewhere], punching, and lemmatization.

5. Preliminary editing for us involves adding to the text, line after line, handwritten symbols and colours that explain what is in fact contained implicitly in the text, but in a way which the average puncher is not expected to understand. For

example, it means things which the addition of special signs show, *inter alia*, to what extent the expression can be attributed to the author himself: thus, we have underlined in green all the phrases of another author quoted by one writer, and in red those phrases in which it is believed that there is a mixture of thought, or of vocabulary, or both, when the author summarizes, outlines, interprets, or refutes the ideas of another.

Preliminary editing also entails identifying and underlining ambiguous punctuation marks, such as the (diabolical) distinction between a hyphen and a dash, underlining the different uses of the full stop, and dealing with what we call "formulae", such as dates and chemical formulae, which must be punched in different codes from those of the words.

6. "Lemmatization" is the grouping within one single paradigm of all the various forms of one and the same word. It also involves the systematization of homographs.

7. I will now give you some facts about the number and persistence of human errors that occurred in our punching. Some preliminary information is necessary to understand these numbers.

7.1. We count as an error a punched card that must be replaced; the decision to punch a new card is made when the first mistake is noticed, but in the rest of this same card there may be other mistakes.

7.2. We call a "text-card" one which is completely punched on 80 columns, using IBM codes, which are combined differently to represent about 100 different symbols of the text.

7.3. Our coding is on three levels, as follows:

7.3.1. Univocal code: for example, if you find *a* in the document, you must press the key *A* on the keyboard.

7.3.2. Simple dicode: for example, you punch the punctuation marks by pressing the number keys; at the beginning of each phrase underlined in green, you punch three times: 3–8, 12–1 and 12–3–8. At the end of the same phrase underlined in green, you punch 3–8, 0–3–8, if no other colour follows; if another colour does follow, you punch the sign proper to it: 3–8, 12–2 and 12–3–8 for red etc.

7.3.3. Complex dicode: when we give the operator more than two rules with which the phrase must be compared. For example: if you come across the Latin word *AB* printed like most other words, you punch it using the codes for "words" (it is in fact a preposition). If, however, it is printed in another way, for example as the geometrical expression *AB*, you punch it as a "formula". If you are undecided, ask the head of the department. The dot is another case in point. It is sometimes used as a decimal point, and so has its own code within the "formulas". It can also be a punctuation mark indicating the end of a sentence, or the end of a group of phrases making up one long single sentence. If you are sure that it is a punctuation mark, punch it as such. The dot will be a punctuation mark when the following word, which would otherwise begin with a small letter, begins with a capital. (This must be checked for each book, for it is not a universal printing convention). But if you think that the following word should always begin with a capital, ask the head of the department. However, the dot is also used as a sort of last letter in a word that has been abbreviated: in such a case use punch 11: in fact, it must be treated in a completely different way. And if the first letter of the following word is a capital and the word is not a proper noun, punch the dot as a punctuation mark: many printers use only one dot for both these functions, i.e. full points in abbreviations and punctuation marks. Do the same if you find two consecutive dots, also if you find three consecutive dots, such as suspension points.

7.4. The figures that I am giving you were obtained through our operational calculation. Each person working in our Centre prepares a punch card for half a day's work, or less. This contains their own identity, the quantity and references of the material used, the mistakes they made while working, and those each one found while checking their own work or that of another. Thus, it is always possible for us to trace each mistake and to find out who made it. We have split our operations into three groups: production, checking, and correcting. The amount of checking and correcting human mistakes increases exponentially, when one increases the amount of the unit.

8. All the work units were completed in the course of the same trimester.

8.1. *Work unit code n. 37:* 1,144 text cards were punched in 8 man-hours. 49 mistakes, i.e. 4.2%, were found and corrected during punching. Another operator using an IBM 056 found a further 21 mistakes, i.e. 1.9%. We then printed all these text cards, and other operators checked them, reading them line after line and comparing them with the original document. They did not find any more mistakes. So, the total stood at 70 mistakes, i.e. 6.1%. However, before you express satisfaction at these results, wait a moment.

8.2. *Work unit code n. 38*: 41,888 text cards were punched in 99 man-hours. 1,041 mistakes were found and corrected during punching, i.e. 9%. Checking by IBM 056 revealed a further 682 mistakes, i.e. 8.8%. Checking the list revealed another 25, i.e. 0.25%. The total of mistakes was 2.021, i.e. 10.05%.

8.3. *Work code n. 35*: 41,888 text cards were punched in 470 man-hours. 2,185 mistakes, i.e. 5%, were found and corrected during punching. The first IBM 056 check revealed a further 682 mistakes, i.e. 1.6%. I had another IBM 056 check done on the same cards, and I was rewarded by the discovery of another 245 mistakes, i.e. 0.5%. These cards were then printed and checked against the original document. Would you believe me if I told you that another 37 mistakes were discovered only at that moment? We totalled up 3,149 mistakes, i.e. 7.18%.

8.4. To get a rough idea, the number of mistakes detected in the last check should be multiplied by at least the number of previous checks.

8.5. In one single operation we achieved a maximum average of 37.6%, which I might call an Olympic record of inaccuracy. It occurred when we punched 13.025 cards of German summaries of biophysics and biochemistry. The reason for so many mistakes was certainly not the German language, for when we punched the Dead Sea Scrolls in Hebrew (even without having Hebrew letters on the keys of the IBM 026), we did not make any more mistakes than when we punched an Italian text.

The German documents, however, were a bad and dark photocopy of a typed text. Moreover, the distinction between a hyphen and a dash was a complicated business because of new factors: the dash was used, as is common in printing, also to indicate a word which begins at the end of one line and finishes at the beginning of the next. The summaries were typed on pre-printed forms, surrounded by a printed line with 2 or 3 small dashes to guide the typist. It sometimes happened that the last letter of a word was typed too close to these lines, and was taken by the operators for a hyphen or a dash.

Lastly, a progressive code number had to be punched on the individual cards (which could not have been done before because the ascending series was discontinuous) and so, every time that a mistake was made, all the following cards had to be punched again up to the interval that followed in the ascending series.

8.6. These changes in the number of mistakes confirmed that it was the reasons for the mistakes which, when checked, improved accuracy.

9. Our empirical classification of the mistakes and the causes of the mistakes began with grouping the mistakes into those due to the document, those due to the operator, and those due to the ambiance and atmosphere of our Centre. The mistakes seem to be a sort of mixing of one set of data with another, perhaps by chance, because of their proximity and the lack of constant attention.

9.1. This fault, when it was repeated, seemed partially to be the result of a lack of motivation or encouragement in human relations within the Centre. It also resulted from the biotype and character of the operator. The accidental interruptions seemed to be the result of temporary emotional situations. Believe me, it is not wise to give a text that is difficult to punch to an operator whose children are doing exams or having an operation. When a bachelor punched Michelle instead of Washington, I realized that there was a 97.4% chance that he had had a row with his fiancée the previous evening ... Accidental distractions are also caused by noise and adverse weather conditions, etc.

9.2. The proximity that we found within a set of data could be of different types. We know that the similarity could be, for example, graphic, phonetic, or semantic. For example, in Hebrew *resh* and *daleth*, *yod* and *waw* are written in a similar way, as are the capitals *alpha*, *delta*, and *lambda* in Greek. The similarities in Hebrew words when pronounced, and as a result vocalized, which do not exist in the orthography of the same words when they are not vocalized, and written badly by the scribes, made us realize that at the monastery of Qumran amongst the lesser Essenes, one man read and dictated, while another listened and wrote.

9.3. Sometimes the similarity lies not within the document, but in some information which the operator already has, as a sort of routine, stored in human computers through the repetition which comes from the environment or from education, or which comes about through nature or by chance.

9.4. For example, you know that many Latin words end in *m*, and only a few in *n*. I had two female operators who, with miraculous consistency, punched *n* instead of final *m*.

9.5. I discovered that they came from the Veneto, where in dialect there is a high percentage of final *-ns,* but never final *-ms.*

9.6. Coming back to graphic similarity, I got some help from the rules of textual criticism established by German and Roman palaeographers of the Bible.

9.7. Thus, there is haplography when the same word is repeated twice in the document, but the scribe writes it only once.

9.8. Conversely, dittography occurs when the word is written only once in the document, but the scribe writes it twice.

9.9. When a word is repeated in two successive lines, and the scribe jumps from the first line to the second where the word is repeated, this is called, *homoarkton* or *homoeoteleuton.*

9.10. After this experience we realized that scientific and psychological research into the rules of human mistakes needed to be carried out.

10. We were thus led to the following conclusions:

10.1. From a linguistic point of view, it will be necessary to undertake a supplementary task: research into the similarity of words. Such research seems to be very interesting, but it requires a lot of time, all the more so when it is added to research into homography, which is already difficult in itself.

10.2. We are certain that the information that we seek exists somewhere. We presume that we shall be able to find something on this topic in the proof-reading departments of printers, newspapers, and post offices.

And this must be improved, because the automation of information processing requires greater capacity, because of the complexity of the coding.

10.3. We hope that Centres of Psychology and Communication will find:

- suitable texts to test people's proof-reading ability,
- programs to train such people,
- the rules of human behaviour regarding mistakes in the preparation of computer input.

Chapter 12 Models of Knowing and Speaking

Roberto Busa S.J.

First published as: Busa, R. 1981. Modelli del conoscere e del parlare. In *Il sapere come rete di modelli – La conoscenza oggi (Convegno Intern. Modena 20–23 Gennaio 1981)*, 235–244. Modena: ediz. Panini.

Editors' note

Busa begins this paper by delivering a series of reflections on how "knowing" can be conceptualized as a network of models, the theme of the conference at which he delivered this paper. He argues that knowing should not be examined in isolation from speaking, for knowing and speaking are found on opposite points of the same continuum. Also, "the expression of man" has a direct relationship with humankind yet cannot exist independently of it and, as such, speaking and knowing can be described as models. To explore the nature of the reciprocity that exists between model and modeller, he invokes the computer program. A program does not contain its programmer but it cannot exist without a programmer and every program bridges two "opposite realities" of "programmer and source program, program and target program". Busa goes on to argue that models of human knowledge can be found on two levels, knowledge that speaks and knowledge that wishes to speak. After examining these models further, he reflects on topics pertinent to them, such as the concept of knowledge itself. In the latter part of the article (and without contextualizing his work with regard to such models) he turns to describe the application areas that his philosophical and lexicological interests were giving rise to: "I am trying to find out if in the vocabulary there is a lexicological variant between the words that recur with regard to any argument and those that are specific to specific arguments". He also reports some of the findings of the lexico-statistical analysis of the vocabulary of the *Index Thomisticus,* which would be published more than a decade later (Busa 1994). This article draws attention to the way that Busa's writings sometimes sat outside formal academic discourse. For example, though he seems to draw on the writings of scholars in areas like linguistics and semiotics to build his arguments he neither acknowledges them nor cites their work.

Works cited

Busa, R. 1994. *Inquisitiones Lexicologicae in Indicem Thomisticum.* A Roberto Busa S.I. latino sermone confectae atque a Philip Barras in anglicum sermonem translatae. 2a ed. emendata auctaque. Gallarate: CAEL.

1. Prof. Betti has just shown us a nice picture: I should like first of all to put a frame around it and then offer some comments.

1. When one starts to talk of artificial labyrinths, one must above all bear in mind that speaking is the fundamental human labyrinth, "naturally artificial", i.e. made with art but by force of nature.

1.1.1. On the basis of the binomial *langue-parole* or *competence-performance* or *code-message* or whatever else you will, there is a correspondence in that knowing and speaking are the two poles of the same operative arch that is expression or expressing oneself. They are necessarily connected because they are the extremes of

J. Nyhan and M. Passarotti (eds.), *One Origin of Digital Humanities*,
https://doi.org/10.1007/978-3-030-18313-4_11

a generative fact, but they are not reciprocal: it is the former that gives rise to the latter and not vice-versa. They are distinct and opposite: the expression of a man is of him and made by him, it does not exist without him, but it is not him. It is his "model", i.e. an image. Every image is something other than that of which it is the image. Every expression creates an irrepressible relationship of alterity between the original and the image.

This is why I cannot agree that the Christian Middle Ages placed God at the centre of the model of the world, just as I do not find Giorgione at the centre of his *La Tempesta*. Likewise, I would not find the name of Giuseppe Verdi at the centre of a formula or a graph that showed me the combinatorial modules of his orchestrations. Every author is not the centre but the father of his own works, and no father is at the centre of his son.

1.1.2. Such concepts are usual in informatics: no program loaded into a machine contains the programmer either at the centre or at the periphery. Yet, nonetheless, every program is an operative arch that inseparably connects two poles, two opposite realities: programmer and source program, program and target program. It is not for nothing that every program is due to its programmer.

In the Middle Ages people were so aware of the binomial thought-expression that they saw an expression expressed by a programmer in every logic incorporated in the structure and behaviour of any event or thing that can be encountered in the great sea of existence. A most arid formula "*compositum causam habet*",[1] means precisely: that every system, every organization, every dynamic cycle are the children of a thought, expressions of a programmer. Not for nothing have the French, who are so refined in precision of expression, rechristened software "*le logiciel*" and hardware "*le materiel*". For every time that I see a program running, I know that it has been well thought-out beforehand by a programmer. And not less but even more so, when one is dealing with programs for probability forecasts or stochastic processes.

1.2. I want to say that the theme of the conference at first sight seemed to place us in a labyrinth: the network of models of knowing. In fact, it has driven us into two labyrinths: that of knowing and that of speaking. Knowing is the father of speaking ... or at least it ought to be ... . And in a scientific domain, knowledge is known only from its expressions, the main one of which is speaking. By scientific domain I mean that of knowledge relative to a community, of acquisitions of experimental data, and of consequent methodologies of construction of knowledge from data.

In reality, therefore, models of human knowledge are to be sought on two levels, that of the knowledge which speaks, expresses itself, and is generative, and that of the knowledge that is expressed with signs, or signified, or spoken, or formalized, which wishes to speak. Between the two there are conspicuous differences, not to say enormous ones.

[1] For instance, in St Thomas, *Summa Theologiae, Prima Pars, quaestio 3, art. 7 corpus.*

1.3. Firstly, speaking is a model of knowing. But knowing too is a model in its turn: and of what else if not of existence? But is existence in its turn a model, an expression? Always? And of what other existence?

1.4. Leaving these runways to philosophy true and proper, we come to the differences of type of model between our two. The "spoken" one, i.e. discourse, spoken or written, is of linear type: the signs, both phonemes and graphemes, follow each other one after another, one at a time, like the bars in a piece of music.

On the other hand, knowledge presents itself to us as co-present and multidimensional. Knowledge is global vision, and—so to speak—continuously instantaneous, of a collection of very many things, co-present in thought, which can correlate each one of these with each of the others, i.e., by taking each one as a centre. These "things" are not only those that have been tested here just now by me, but they are past, present, and future, visible or invisible, known and even more so if unknown. I mean, with evident preferences for the things that are unknown—as in all research into why—or not yet in existence, as in the planning for realization, or in the need for guarantees, or things that exist no longer, such as, for example, in regrets, in the unfashionable, and in history.

1.5. Furthermore, discourse is finite, defined, closed, but speaking knowledge is open, generative, dynamic.

1.5.1. We can analyse written or spoken words, but as for those with which the mind might express itself, not even he who possesses them could ever count them. Which of us could say how many words he knows? Grammars state that superlatives exist only for adjectives: but if an author were to write "the swordest of swords"[2] we should see not an error but a likeable expressive stylistic construction.[3] Knowledge is an overpowering and creative activity, which attacks and seizes things that are not it, such as sounds, graphemes, colours, spaces, volumes etc., and from them makes expressions of itself: it succeeds in communicating itself to others through things that are neither of the two interlocutors. This instinctive power of symbolizing, of ordering other things to act as a "sign" —signs of one thinker through another thinker, signs which, like things, are neither the one nor the other—is a reality which, let us say, is also mysterious, which presents operative dimensions of a nature that is other than and superior to that of physicochemical activities.

1.5.2. Knowledge always transcends the bounds of the immediate, i.e. of the present touched by hand, avid as it is to go on safari in the world of the new, the possible, the thinkable, of that which no longer exists or has not yet come into being.

---

[2] [Editors' note] In the original the term is given as *un cavallissimo*, which is the superlative form of the noun *cavallo*, i.e. an unusual formation.

[3] [Translator's note] cf. Ratherius of Verona, "Deus est spiritissimus omnium spirituum".

1.5.3. Knowledge is not only, as in a computer, memorized information. It is also unbridled curiosity, an inventive creative capacity, as well as an unappeasable judge because of the demands of critical logic.

1.5.4. The speaking labyrinth is the most intricate, the most labyrinth-like of all labyrinths. It is that medley of operative thrusts, of springs released that everyone has within him somewhere, but no-one knows exactly where. An encyclopaedia is nothing but the inventory of one of the rooms of its trophies.

1.5.5. Nonetheless it is not chaos. In fact, paradoxically it is also a centripetal force that sucks back towards the unity of knowledge, if nothing else by making it regret when it seemed lost. It would seem in fact—equally paradoxically—that it behaves in this way so that the great sea of existence is its light, life, and dynamism. None of us is able to say with other words what existence is, but our vocabulary contains it as that unique verb "to be", which is the most used word: an indication of our awareness that it is the beginning and the centre of unity.

2. A few exits on this line of comparisons. Firstly, to continue the analogy between thought expressing itself and a computer, even in thought, as well as the peripheral unities of input and output, there is a central elaborative unity.

2.1.1. Now, in thought that expresses itself (it does not always do so, and not all thoughts are expressed) the central unity has, among other things, the enormous function of personalizing the information received from another human being, i.e. of examining it and then, if merited, of making it its own. It is one thing to say "Jack tells me that this is how things are", and quite another to say "This is how things are". The information given to a person is not a physical recording, physically and mechanically imposed from outside; it is not like inserting a magnetic tape into a unity of a text, but it is like ringing a doorbell and asking "Do you want to buy this?"

2.1.2. Information should not be given to a person merely to make him memorize "everything already made and ready for use", but, as a matter of principle, to give him a start because it is he who must understand and proceed.

2.1.3. To give information is not to illuminate dark areas and still less is it to give help to the poor, but it is like putting a slide between the light and lens of a microscope. Only when we talk to someone who understands do we have a dialogue and a transfer of information when the dialogue is not merely communication of life. Understanding is a personal exercise of light that is active, aware, critical, and perhaps suspicious. Not for nothing is there an active verb in Latin that means "to seize". Understanding is something very different from the passive and succubus-like reception of magnetized bits on a coat of ferrous oxide.

2.2.1. There are also two exits on the opposite labyrinth, that of discourse. First, there are two opposing situations that are quite distinct and need to be studied separately: face to face dialogue and the written text. In the former the reciprocal communication of thought has many other very strong supports beyond the verbal signs in the strict sense: signs such as look, mime, tone of voice etc. Not for nothing are lexicon and grammar much less in evidence in languages that still lack literature, i.e. in those that exist only in dialogue and conversations.

2.2.2. In my paper (not published here) with regard to Prof. Romano's excellent opening speech, I referred to a paradox of mine, that of the "diffuse centre", and I connected it with a sentence of a pseudo Hermes Trismegistus which rebounded down the centuries from Alan of Lille to St Thomas,[4] to Rabelais, Giordano Bruno, Pascal, Voltaire ... "*Deus est sphaera infinita cuius centrum est ubique, circumferentia nusquam*" (God is an infinite sphere whose centre is everywhere and circumference nowhere).[5] I should like to take this idea of a diffuse centre up again, precisely with regard to the model of written discourse.

2.2.2.1. I see the linear progression of a text as rigorously "uncentred" at the level of the system of the signs that are verbal or made by the expressions. Every string of graphemes is correlated in time and space only with the two contiguous ones, that which precedes and that which follows.

2.2.2.2. It is necessary to go back to the semantic levels, i.e. to those that are signified, that is to say still to rely on the discourse, either that which expresses itself or the mental one, to find the centres, such as the verb or subject or predicate in a sentence (according to the perspective taken), or the argument in a court ruling. These are true and proper centres, because they exercise an influence not only on the contiguous words, but also on those that are at points far distant from the net. At the moment I am concerned precisely with a diffuse centre in the semantic levels of discourse in a seminar of lexicology that I am directing at the Gregorian University in Rome.

2.2.2.3. I am trying to find out if in the vocabulary there is a lexicological variant between the words that recur with regard to any argument and those that are specific to specific arguments. In the delightful paper that Prof. Bouligand has read to us on the model of the shell of crabs, the presence of the word *calcite* is demanded by the argument. It is certain that in other arguments that word would not appear at all. But words such as disposition, organization, structure, all and some, sooner or later, one and many etc. certainly did occur in this text, but that was because they occurred in every and any other text. So, it is precisely these that I mean when I talk of being a diffuse centre: centre, because they are supporting structures, diffuse, because they are everywhere.

2.2.2.4. It goes without saying that I place among these also, but not only, those which the French charmingly call *mots outils* (I should be annoyed if anyone were to call them *mots vides*, for only words without meaning are empty): prepositions,

[4] St Thomas, *De Ver., quaestio 2, art. 3, arg. 11.*

[5] Liber XXIV Philosophorum, ed. Cl. Baeumker, Muenster 1927, *Beitraege* etc. XXV, 1–2, pp. 207–214. See also E. Gibson, *Philosophy in the Middle Ages*, Italian translation *La filosofia nel Medioevo*, La Nuova Italia, Florence 1973, p. 378.

conjunctions, pronouns, articles, auxiliary and modal verbs, numerals, adverbs of time, motion, and place, and what else? I am of the opinion that all these words should unquestionably be categorized with those words that are common to every argument. And if I choose to categorize the first words as functional, it is amongst those words which are absolute and constant that those words which are general or common should be separated from those which are specific or classified.

2.2.2.5. To my eyes such a lexicological distinction extends to the distance in two directions. The first is that lexicography is very different for the two types: it is one thing to define words such as "house" and "horse", quite another to define words such as "dimension" and "movement". As a consequence, we should clearly distinguish two portions of the lexicon, both in themselves and in the model of their diachronic evolution. The lexicon of specific words would be just the dictionary or vocabulary, while that of common words would, at the same time, be true and proper philosophy.

2.2.2.6. And this is the other direction, in my eyes, to which this idea of mine of a diffuse centre of language opens up, an idea which at the moment is just a hypothesis for research. There I catch a glimpse of the expression of that first, natural, and personal logic of existence, which is incorporated in the central unity of our knowledge and understanding.

3. I come at last to the contents of the paper given by Prof. Betti. To individuate the model of the net or nets of existing correlations in the common knowledge of today, implies individuating—on a selected sample such as the *Enciclopedia of Einaudi*—the nodes of mental interests from which the correlations spread out, or rather, the points that are correlated there.

3.1.1. One of the duties of "artificial intelligence" is that of making it clear, if nothing else, if it is true that one day it will be possible to get there or not. The problem, for me, can be reduced to that of knowing whether we shall be ever be able to formalize the global meaning of a document in very much less than all the words that compose it.

3.1.2. *Inter alia*, I have made an inventory also of a file of 10 and a half million words, the *Opera Omnia* of St Thomas Aquinas, and 61 works by other mediaeval authors. The results are that:

a) the frequencies of the words, whether single or trinomial, tend to be like a mountain peak: very few words and very few combinations of words are very frequent, while very many are the least frequent;
b) nonetheless, the most frequent are precisely those functional words – *v. supra* – which cannot function as key-words of the specific content of any text. Here follow some more precise data.

3.1.3. My linguistic universe, in this file, consists of 8,767,855 words of St Thomas, plus 1,864,125 words of other authors. There emerged 133,752 graphic forms, which further developed into 147,088 semantically different forms, a figure which in reality could be much greater for various reasons. So, not counting proper names, numerals and abbreviations, amongst these 150,000 in St Thomas there are only 880 forms of words with a frequency higher than 999: these amount to 5,872,138 words of the two texts, i.e. 66.97%.

Furthermore, amongst these there are only 85 words with a frequency higher than 9,999: these amount to 3,675,575 words of St Thomas, i.e. 41.92%. Everything considered, absolutely the most frequent form is *et*, with a frequency of 295,593. This is followed by functional words: the first non-functional word that is met there is *homo*, with a frequency of 15,879; then there are another 7 functional words, then *peccatum* with a frequency of 13,207.

3.1.4. At the level of lemmas, of the 20,173 "most compressed" lemmas of the whole, only 128 lemmas in St Thomas have a frequency higher than 9,999 and total 5,285,963 words, i.e. 60.26%.

3.1.5. Here follows more data, in tabular form, which, however, does not include the non-lemmatized forms in the addition. Of the words in St Thomas that are neither proper names nor special words:

	Lemmas	Forms used	In the whole of St Thomas	In other authors
nouns, adjectives and pronouns	11,366	46,417	3,205,264	724,754
verbs	3,796	70,350	2,060,649	419,810
invariables	362	441	2,481,895	506,106

Table 12.1. Words other than proper names and special words in the works of St Thomas

3.1.6. Also, of the forms in St Thomas, always excluding the ones mentioned above, 30.22%, i.e. exactly 29,637, are *hapax legomena*, as are 2,229, i.e. 16.9%, of the lemmas.

3.1.7. From the complete file of the works of St Thomas I have extracted a file of 5,962,981 trinomials. Amongst these there are 70,430 different key-words, each one accompanied by a preceding and a following word, when there was no heavy punctuation in between. I put them in alphabetical order. The result was that the non-identical trinomials, i.e. those in which at least one word was different, amount to 3,360,425, i.e. 56.35%.

Of these:

- only 117 (0.0003%) recur more than 999 times, and are all functional words, up to a maximum of 15,556 (*et ideo*), followed by 6,870 (*et sic*); of these only 68, which had a maximum frequency of 4,851 (*ad primum ergo*), are trinomials: the others are binomials, because of heavy punctuation next to the key-word.

- then 2,092 (0.0622%) have a frequency between 100 and 1,000.
- another 545,984 (16.24%) have a frequency between 2 and 99
- finally, 2,758,653 (82.09%) recur only once.

3.1.8. These figures speak for themselves. When I enumerated the labyrinth of speech from this perspective, I did not expect such a fragmentation of combinatorial varieties, or rather I expected many more trinomials identically repeated. I put this on record as a proof of the richness of creative freedom that knowledge has to express itself. However, since I knew that there are 3,375 plus 12 noughts, i.e. 3,375,000 billion, combinable trinomials, I deduced that in its use of freedom in its combinatorial choices of words, knowledge shows itself to be restrained by a network of parameters of possible or impossible or in compossible logico-semantic correlations.

3.2. Therefore, I must agree with Prof. Betti, when he adopts a machine-human interaction as his methodology.

3.2.1. By "machine" I mean the computation, which can or could be computerized, of how many times a given expression, chosen as characteristic, recurs in the context of another given expression. By "human" I mean the capacity that human beings have of seizing and reconstructing, by reading, the global and comprehensive meaning of the text, and of then determining its weight, i.e. the proximity to the argument that each of these expressions has when it is seen in correlation with it. For a programmer – after having progressively accumulated such data – it will be a mere game to computerize them, until he is able to visualize their network in formulae, in graphs, or perhaps on a colour video.

3.2.2. This network of points of concentration of cultural interests is a most genial new form of index, analytical or real or by subjects - whatever we call it.

Prof. Betti takes what is written in the *Enciclopedia Einaudi* as representative and descriptive samples of the communitary reality which is every science. From these, with the circumstantial methods of lexicography, he goes back to the whole of the inner knowledge of the authors of what is written, which is understood to be representative of common and current knowledge.

3.2.3. It is obvious that all this is according to a synchronic perspective. Such a network, if valid, will be adopted as a model of knowledge today. The clear fact of the evolution of every model of knowledge lies behind every history of ideas, of science, and of culture.

3.2.4. This evolution is a complex phenomenon within which there are corresponding zones of involution. For example, the disproportions that the *Enciclopedia Einaudi* reveals in religious experiences are to be understood as nonetheless descriptive of the fact that, in the culture that is widespread today, the experience of prayer, dialogue with the one Father of all, is in a clear phase of involution, because in corresponding increase is the concept of "all without a beginning", that is, without a common beginning other than the matter of which we are made, and therefore, in the final analysis, without a possible normative beginning other than the state.

3.3. A last reflection: what type of words can signify the points of concentration of the actual mental interests? It could be that in reality they are only the 600 title words of the article of the encyclopaedia. If that were so, the selection would have already been wrong, and the only thing to do would be to examine the words.

3.3.1. In any case, such words are not to be treated rigorously as descriptors or the key-words of a *thesaurus* in a search for documents, i.e. those with which, using for example the pack of the programmes STAIRS-IBM, I retrieve from a database the list or even the text of the documents that contain them.

3.3.2. Second: such words, having to signify mental interests, indicate the presence of "values" that push one to use them repeatedly, by connecting much of the rest to each one of these, so much so as finally to make of them the centres and nodes of correlations.

3.3.3. None of the functional words is certainly among them. And there will be those words, at least not prevalently and not even in large numbers, that signify entities or single events, such as proper names, dates, and acronyms.

Certainly not many common words in the strictest sense (*v. supra* 2.2.2.) will be present, but there will be many words that are common in more ... common and larger meanings, such as liberty, democracy, sociability/sociality, well-being. Among these, many words with defined boundaries will be lacking, because they will not signify connections or elementary values, but rather groups of complex and pluridimensional wholes.

4. I conclude by saying that the geniality and novelty of this idea of an encyclopaedia of knowledge and of the visualization of its indices as networks of points of concentration of mental interests has filled me with admiration, although I am not qualified to assess the greater part of its contents.

Chapter 13 Thirty Years of Informatics on Texts: at What Point are We? What Opportunities for Research?

Roberto Busa S.J.

First published as: Busa, R. 1983. Trent'anni d'informatica su testi: a che punto siamo? Quali spazi aperti alla ricerca? In *Conv. su L'Univ. e l'evoluzione delle Tecnologie Informatiche - Milano 14–16 Marzo 1983, Atti,* 2 voll., 7.1–7.4. Milano: CILEA.

Editors' note

In this text, Busa bemoans the state of the art of linguistic informatics as it looked to him in 1983. He argues that the field had made little progress in the previous thirty years and that its focus had not shifted from individual words. Research on areas like automatic lemmatization, on groups of words with similar meanings and on automatic translation remained embryonic. As a result, little real progress in the computational formalization of language could be noticed. The field showed a marked lack of imagination and was not fulfilling its potential of transforming into an industry. Busa suggested some research directions that the field might pursue and argued that it required new types of "philological-informatic programming" to move forward.

This paper is interesting because it shows Busa interlacing old ideas (or ideas he had often ruminated on in earlier publications) in new ways. For the first time in the articles gathered for this collection, he seems to consider the idea that the limitations of computing speak not only to spiritual concerns, and questions about what it means to be human. He also seems to suggest that completely new kinds of computing are needed to formalize language: "We must find new methodological techniques and new strategies of research … which cut out the fabric of mathematical-statistical calculations according to the meta-mathematical measurements of the intuition of the whole, an inner force that wishes for proportion, harmony, beauty, meaning and value".

This paper is also interesting for what it does not say. For example, Busa writes approvingly of those who have questioned whether word frequency counts are useful for interpreting text. Yet, his own research placed a not insignificant emphasis on word frequencies and he does not attempt to reconcile this with the above observation. To what extent is he acknowledging some of the limits of his own work in this article?

Although critical of the progress made in linguistic informatics, Busa does single out some of the contributions that the incorporation of the computer has made to the work of the philologist. For example, what he feels was the philologist's over-reliance on the lemma has been broken as the computer requires lemmas and forms to be studied together. The computer has also drawn attention to the interplay of morphology and syntax in generating different kinds of meaning and he feels that this is not widely appreciated.

1. *Three linguistic informatics: purely structural, documentary, lexicological.*
2. *Two first elementary linguistic problems from databases.*
3. *Linguistic informatics, after its first qualitative leaps, is producing little more than just indices and concordances of individual words;*

J. Nyhan and M. Passarotti (eds.), *One Origin of Digital Humanities*,
https://doi.org/10.1007/978-3-030-18313-4_12

4. *because of insufficient creativity of methods of analysis of the whole, commensurate with the proper nature of the linguistic phenomenon;*
5. *and not as if human expression were incapable of being formalized and computed.*
6. *Of the desired new types of philological-informatic programming;*
7. *previous block diagrams could be sent out from competent centres;*
8. *in fact, in linguistic informatics opportunities for research are incredibly wide.*

8.1.-8.4. Examples of this from a list of subjects for degree dissertations.

9. 0.-9.2. Conclusions.

The nature of the subject obliges me to tell you just what is known. I will give only the type of procedures and not the geographical and historical distribution of people and publications.

I will describe the situation just from a qualitative, or, to be more exact, typological, point of view of the methods and objectives of the research. I will deliberately use language that is common, stripped of all the specialist embellishments used in our journals, well known to readers, apart from the general ones, such as *Informatica, Informatica e Documentazione, Linguistica Computazionale, Computers in the Humanities*, ALLC Journal, *American Journal of Computational Linguistics*, and *The Prague Bulletin of Mathematical Linguistics*.

I will speak about the use of a digital computer in the production of printed texts which contain arguments in words and not just numerical calculations. You will understand how many and what sort of areas I have omitted.

1. I distinguish three linguistic informatics: the purely structural, the documentary, and that which is properly called linguistic, lexicological, or textual. By structural I mean all research into signs whose meaning, granted that they have one, is completely ignored, such as the undeciphered alphabets of unknown languages, or encrypted texts, which is cryptography. By documentary I mean all informatics that point to or are equivalent to any form of a database, i.e. that which seeks all documents that contain one or more keywords. By linguistic, or lexicological, or textual, I mean every use of the computer that analyses in written arguments the signs as signifiers, i.e. relating them to meanings or concepts.

To get an inkling of what I mean by that, it will be enough to try to define what differences there are between musical and verbal expression, and between verbal and figurative, gestural, and mimetic. Such linguistic analyses have two faces: one looks towards the description of the models based on which a text may be produced (and here I insert computational linguistics); the other looks towards the lexicological inventory of linguistic typologies that emerge in texts already produced (and here I insert statistical linguistics).

I touch in passing on the contrast (which in other ways involves informatics in general) between the informatics of the international networks of databases and that of personal computers: the contrast between giants and dwarves. We spoke about personal computers at the Inter-University Day on 26–27 November last year, a conference which Prof. Gc. Bolognesi of the Catholic University and Prof.

A. Grilli of the State University organized and which I chaired. The personal computer, by whatever name it is called, is to be considered as the typewriter of today and of the future, and I fervently recommend that universities or university services install facilities so that students may learn its use and functions. Anyone who is preparing to work with a pen must be made aware that the mechanical typewriter is now an exhibit in museums of science and technology. Perhaps this might tempt students to use their grants to buy a personal computer before buying a car.

2. Documentary informatics meets the need for a social service of information that in the electronic era is one of the irrepressible developments of telecommunication. Its functions are complex, more in terms of technology than of concept. But every time the data is sentences instead of numbers, one is faced with two typical linguistic phenomena, which I call the taxonomy and synonymy of words. Let me explain by examples. Suppose that I want to know the decisions of the courts in disputes about domestic animals between owners of apartments in a block of flats. It will not be enough for me to look up in a database the words "domestic animal". Indeed, the more general meaning of the words is present in my ideas about "feline, canine, winged creatures, canary, cat", but what makes the words "domestic animal" on computer discs apply equally to winged creatures, cats, and canaries?

What I mean by synonymy is when one and the same concept–for the most part one of those that are more widely and commonly experienced–is expressed by many different words: habitation, accommodation, apartment, house, domicile, residence. How can the computer be instructed to produce them all when I give it only one as a key?

3. On the contrary, linguistic informatics, firstly still operates only within the areas of pure science; secondly, it is still re-cycling without superseding the qualitative leaps forward made more than thirty years ago. The first gigantic leap in terms of quality was the integral completeness of the analysis of the texts, in place of the samplings that for centuries were only available to philologists. Moreover, the computer allows and indeed requires that the lexicological system is put together on two levels that have developed separately, that of the forms and that of the lemmas, while traditional philology jumped straight from the text to the lemma. Finally, the computer compelled us to be aware of the difference of the levels of meaning expressed by the morphology of a word and those added and varied by the syntax. Unfortunately, the awareness of this distinction, in my opinion, is still not sufficiently widespread.

As a result, publications of indices, tables, and computerized concordances are innumerable, but if we look at the types of formulas with which they are produced, we find, firstly, that still there has been no general progression beyond the morpho-lexical categorization of the individual words, i.e. taken one by one. Secondly, that the fierce attack on automatic lemmatization, syntactic-semantic analysis and syntactical-logical analysis has hardly begun. For example, have you ever come across computerized tables and concordances in which the programme automatically re-unites names with their respective surnames? Or extracts the compound forms of verbs? Or connects pronouns with the nouns that they represent? Or marks the implied meanings? And I am not asking if it distinguishes the proper use of a word from its metaphorical one!

What I mean is: we are still not able to elaborate smoothly on a computer groups of words, or rather words that are semantically linked and closely related. Still less are we able to elaborate the global meaning of a sentence, and still less that of groups of sentences. Automatic indexing and automatic abstracting are still dreams, and what little automatic translation is being carried out flows in the channel of contemporary technical-scientific writings. Yet, nonetheless, it is clear that linguistics would explode into an enormous industry of information if only these difficulties could be mastered.

4. Why? In two recent conferences (Liege, LASLA, 18–21 Nov. 1981 and Montreal, Chum, 29 Apr. – 2 May 1982) I tried to list some of the characteristics of the linguistic phenomenon that resist the attack of informatics. I separated the steps of the semanticity of words into proper names and common words; amongst these, the deixis, numbers, names of entire objects, their parts and grouping, and lastly all the remaining words. Here too I touched on four of the six linguistic parameters: taxonomy, synonymy, metaphor, vicariance, syntax and discourse. By syntax I mean the fact that in a sentence the individual words are like integrating parts of a structure and have functions that vary according to the structure. The same word in one place is a noun and in another an adjective; in one place it is the subject and in another the object, even without mentioning cases of the type "*su la porta di casa*" and "*la porta di casa in casa*".[1] It follows that, as in a building, the specific structure of the whole (that which is due to the architect) is not the mere sum of the materials of which it is composed. So, in speech there is one global style and one global content that are not the mere sum of the words of which they are composed: meaning and style would change if the same identical words were arranged differently.

Then, from the combination of the progressive components of expression (sentence, passage, discourse), two remarkable characteristics emerge. Expression is, above all, an operative arc whose two poles are not symmetrical and homogeneous. The discourse produced is linear and closed, but that which is generative is multi-dimensional, co-present, multi-centric (or rather, paradoxically, with a centre spread out everywhere). In the second place, every act of expressing oneself is creative freedom within the confined spaces of the thinkable, canalized only by the logics of existence.

Like it or not, the moment comes for everyone when they realize that, in linguistic studies, the wardrobe of physico-mathematical methods lasts only up to a very precise limit, and that in words, the face of the sign is like a physical entity. Beyond that, forces of another logical and ontological category intervene on the side of semantics. I assign these to the dominant intuition of the whole in key are-

[1] [Editors' note] In the first example, the word "*porta*" is a noun ("door"). In the second, it is a verb ("to bring").

as of harmony, beauty, and love; of the whole, beautiful, true, valid, solid, sure and good for ever. Speech is thus essentially a vital expression, of an affective and artistic type: it is subtended not just by the scale of the elements, not just by the laws of thermodynamics and not by mathematical analysis alone.

5. And yet, speech, if only by its very existence, shows that everything can be formalized, i.e. can be expressed through quantifiable physical signs. In fact, what else are words? Every thought is capable of representation. Totality, safety, truth, beauty, love, have always been expressed, spoken, written and sung. … Therefore, the problem can be reduced to asking how one can pass from the formalizations of our thoughts, which is what every written text is, to the algorithmic formalizations of computer programmes.

If, therefore, I share the wise reservations of so many people about the interpretative value of word frequencies, I do so not as a matter of principle, as if I excluded the possibility that frequencies could ever have any importance in semantics, but because I consider that, made as they have been up to now, they certainly have little. If, therefore, in linguistics the computer still cannot quickly recycle repetitive operations, as is done at office counters, there are two points that I wish to underline. One is that we must study what sort of additives will be necessary when linguistic analyses are programmed. … We must find new methodological techniques and new strategies of research which aim at finding and analysing all types and levels of the semanticity of written discourse, and which cut out the fabric of mathematical-statistical calculations according to the meta-mathematical measurements of the intuition of the whole, an inner force that wishes for proportion, harmony, beauty, meaning and value.

The second point is that it is obvious that new valid strategies of research cannot be created from nothing nor derived by mere deduction from hypotheses or dogmas. They can be derived above all, if not only, from many extensive patient and accurate analytical inductions from texts. The more, the better.

6. With that I formulate the first conclusion. For thirty years linguistic informatics has been shuffling about on the same spot, not because of poverty of memory or the functions of the computer, but because of the poverty of our imagination to create formalizations appropriate to philology. Computers are simply our expressions, no more and no less than our words are: why should we not have the wit to think up new instruments of programming? Is it really impossible to overcome the very simple misunderstanding of considering numerical calculation to be the unique, one and only, complete, self-sufficient instrument for acquiring knowledge?

7. As a second conclusion I express the wish that someone would seek to individuate, delineate, plan and publish some, for want of a better word, block diagrams of research procedures, which would achieve the formalization of the semantics of groups of words. In economics they are even exploring what the market will be like in ten- or twenty-years' time. It would be splendid if something of the sort could be offered to philologists, in terms of starting-points, directions, lines of de-

velopment and methods of research. For me it seems obvious to say that the universities could entrust such an enterprise to those private bodies (such as associations or learned journals) or public ones that co-ordinate respective interested sectors.

8. The third conclusion is that even in computational philology there are many vast spaces open to research. It is good to be able to say to young people that there is still so much to do, and that there is a gigantic list of dissertation titles to choose from. Allow me to mention some of them.

8.1. Of the vocabulary of any language or author, a lexicological analysis that was above all general could be compiled. It would group each and every word without exception according to its semantic type: in grammars, only samplings are to be found.

Then detailed studies could be carried out, always on an individual vocabulary, to obtain, for example, a complete and precise list of words to be considered grammatical or functional or "empty", distinguished from those to be considered as constant or "full". Again, to identify all the words that can be arranged (not all can) in hierarchical pyramids, from the more universal to the more specific, and then to identify all those that signify parts of structures that are signified by another and proper word; and, from another point of view, to separate all the words that can be used on any topic from those that are specific to specific topics. Do you not think that documents of this sort would be useful for informatics? I mean that our understanding of the lexicon has to be completely remade, not to say made. Before the computer was invented it does not seem to have been very easy even to extract from a linguistic universe a complete list of all the words that there are in it …

8.2. Another two projects for research would be a basic dictionary and a dictionary of synonyms. All that is needed to remedy previous failures of basic dictionaries is further reading to organize them not only according to the frequencies of single words, but also:

a) according to frequencies of structures of associated words, and
b) according to the aforementioned distinction between common words in any sort of discourse and specific words in specific subjects. A dictionary of synonyms is only a grouping of words according to their "notional field", that which brings together all the different words that express one and the same, or nearly the same, concept. But perhaps we do not know whether, or where, there exists a general scientific and computerized method to produce a dictionary of synonyms that operates beyond intuitive empirical procedures.

8.3. The linguistic facts that we know merely by hearing them spoken are innumerable … For example, I still have not seen treatments that are inductively documented of the phenomenon of homography, nor of that of metaphor, nor of the vicariance of pronouns, for example, nor of the deictics, nor of implied words, those that are expressed by not expressing them. Research into all of these matters could the subjects of many dissertations. The few students of artificial intelligence, or of formalization of discourse that deals with presuppositions both logical and pragmatic, construct a theory based on examples, but I do not know of anyone who has gathered the whole list of cases in its entirety in a natural text. Useful informatics research is always possible, both for semantically profound values such as the use of adjectives or verbs for nouns, or of nouns for adjectives, and for more superficial values such as those of punctuation or capitalization.

I think that there is not one paragraph in our grammar books that, with one or more inductive researches on the computer, could not be finally documented and re-systematized in such a way as to be capable of being inserted into computer lists without causing too many unpleasant surprises. If it is true that it is usage which makes language, grammar *per se* must be employed for censuses of usage and not just reproduced without alteration of form.

Conjunctions, modal verbs, auxiliary verbs, logical functions, clauses, semantic correlations (for example, relations of noun and adjective, preposition and verb, and of a preposition and its object) and all the rest, when gathered and analysed by the computer in texts of sufficient size, would not only offer a researcher a robust and healthy formative training, but would give all of us precious factual information for programming.

8.4. Among the disciplines which, in my opinion (as I said at the conference at Gallarate in June 1981, whose proceedings have just been published, *Global Linguistic Statistical Methods*, Ateneo, Rome 1982 pp. 111)[2] are yet to be constructed almost in their entirety, i.e. in which almost everything yet remains to be done, I cannot forbear to mention linguistic statistics. If it is true that it makes sense to speak of global statistics of a text, i.e. statistics that progressively summarize the quantitative behaviour of every single element, starting from each single homogeneous portion and arriving at the whole as at an entirety, and if it were true that a text of 100,000 words would be enough for a dissertation, then even the magnetic tapes of my *Index Thomisticus* alone would represent an input already available for more than a hundred dissertations: there are ten and a half million record-words of 150 bytes, of which 22 are for the word and the others are for all the qualifications of location and typologies. Global statistics and mathematical formulation of a text have the same value: but one must keep in mind that they bifurcate into the line of the style and into that of the contents or message.

9. Precisely because language is creative freedom, research on a subject which is performed by the computer on a text or on an author, must be repeated on the same subject on many other texts and other authors. Anyone who studies only Wagner would never know the musical possibilities of Rossini and Verdi. The spaces of artistic and spiritual creativity are the realm of the diversities that enrich it, and they are open to that "ever more and always different" which thermodynamics forbids to the transformations of physical energy. Moreover, it goes with-

[2] [Editor's note] The publication referred to is Busa, R. Ed. 1982. *Global Linguistic Statistical Methods to locate style identities - Proceedings of an International Seminar, Gallarate, June 5–7 1981.* (Less. Intell. Europeo XXIX). Roma: Ateneo.

out saying that where the measurements cannot be more than probabilistic, it is always prudent to multiply the surveys. Therefore, if my list of subjects for dissertations only came to 200, if multiplied by 100 works or authors it would come to 20,000, and if by 1,000, it would come to 200,000. How many years would this last for?

9.1. In conclusion, I am sorry that all that happens is the production of concordances, which, alas, too often are not even lemmatized, and which then nobody studies . . . I heartily recommend that these be elaborated as input of further progressive linguistic research. If five firms were called upon to make a road through a virgin forest, it would not be very intelligent for them to make five separate parallel ones, each firm making its own . . .

9.2. I have been using the following metaphor for years: let us not erect a kilometre of algorithms on a centimetre of inductive base, but let us first of all erect on a base of a kilometre only the first centimetre, and then another one on top of this and so on. For this reason, in recent conferences abroad I have also expressed the wish that there be added to the plans I have suggested here this evening a list for publication that gives the titles of dissertations, those announced as possible, those already begun, and those that have been completed. You will be glad to know that I have now finished.

Chapter 14 The Complete Works of St Thomas Aquinas on CD-ROM with Hypertexts

Roberto Busa S.J.

First published as: Busa, R. 1990. Tutto S. Tommaso d'Aquino su CD-ROM con hypertext. *Informatica* 3/90 XX(3), Lug.-Sett., 273–276. AICA: Milano.

Editors' note

Here Busa reports on a seminar that was held in Brescia, Italy, in 1990, on the nature of databases, CD-ROMs and books. He begins with details of who attended and supported the seminar and proceeds with an overview of the approach that was devised by De Antonelli and others to transfer the *Index Thomisticus* from magnetic tape to CD-ROM. Busa then discusses the nature of the interrelationships that exist between books, databases and hypertexts, drawing on examples from the *Index Thomisticus*. In his subsequent discussion of the *Index Thomisticus* as hypertext, he describes it as being of a "completely different" and superior nature to other hypertexts. His categorization of the term "hypertext" as being new, and the limited definition that he gives of it, indicates that he probably was not familiar with the work of Nelson (for example, 1965) or the literary hypertexts of Malloy and others (see Malloy 1986; and also Grigar and Moulthrop 2015). Other aspects of this article have much currency though, for example, when Busa warns of the dangers of using historically inappropriate language in what would now be called data science and text mining research, and even in simple searches of texts: "A very bad use of the CD-ROM of St Thomas, which I deprecate and fear, is that of seeking Italian words with Latinized endings which correspond to one's concepts of today, without first having tested their semantic adequacy".

Works cited

Nelson, T.H. 1965. A File Structure for the Complex, the Changing, and the Indeterminate. In *ACM '65 Proceedings of the 1965 20th Nat. Conf*, 84–100. New York: ACM.

Grigar, D. and Moulthrop, S. 2015. *Pathfinders: Documenting the Experience of Early Digital Literature*. http://scalar.usc.edu/works/pathfinders. (Accessed 04/03/2019).

Malloy, J., 1986. *Uncle Roger. Whole Earth 'Lectronic Link*. Republished at http://www.well.com/user/jmalloy/uncleroger/partytop.html. (Accessed 04/03/2019).

The seminar entitled "The electronic book comes of age. The Opera Omnia of St Thomas Aquinas now accessible on CD-ROM" was held at the Brescia campus of the Catholic University on the 15th and 16th of May 1990. There were two round tables on the following subjects:

- CD-ROM—Database—Hypertexts: what they are and what they do;
- Database with Hypertexts on the one side, Book on the other: peace or war?

Proceedings were opened by Prof. Carlo Banfi, Dean of the Faculty of Mathematics at the University. The Brescia campus of the Catholic University and EBIS (*Ente Bresciano Istruzione Superiore* = Brescia Council for Higher Education) were represented by L. Schiavini and Gfr. Porcelli, IBM-Italia by S. Pezzani, the

J. Nyhan and M. Passarotti (eds.), *One Origin of Digital Humanities*,
https://doi.org/10.1007/978-3-030-18313-4_13

publishing house "La Scuola" of Brescia by M. Bernacchia, the "*Giornale di Brescia*" by G.B. Lanzani, and the "Soroptimist Club" by G. Conte Archetti. L. Dadda for the CAEL (Association for Hermeneutic and Lexicological Analysis), G. Mondinelli for the TREND of Brescia, E. Pentiraro of the "Editel" of Milan, and R. Betti and P. Slocovich for the SII of Milan, related how they had transferred the Thomist texts onto CD-ROM (as was announced on 28th Nov. last year at a ceremony at the Catholic University).

The meeting was also honoured by the patronage of the AICA, of the Soroptimist Club of Brescia, and the Association of the Former Students of the Istituto Arici. The subjects discussed aroused much lively interest; in fact, they connected past and future, but above all they set the marvels of technological evolution firmly within eternal humanistic and spiritual values.

The 118 works of St Thomas plus the 61 works of other mediaeval Latin authors filled 14 magnetic tapes of 2400 feet, at a density of 6250 with 10,631,988 records, one for every single word. Every record consists of 152 bytes, of which 130 contain information relevant to the word, and they constitute a true and proper hypertext of the discourse, the fruit of the work begun in 1946 and finished in 1980 with the publication of the 70,000 pages in 56 large volumes of the *Index Thomisticus*. Thus, the whole consists of 1,616,062,176 bytes. These were first of all compacted into 117,510,144 bytes in such a way that no information was lost, plus the 608,384 bytes of the "indices". This was achieved by first of all suppressing the blanks and replacing the most frequent words with codes of one single byte, according to the criteria established by the codifier D.A. Huffmann, and progressively replacing the less frequent words with codes of more bytes, and finally by loading everything with "LT (lexicographic tree) structures", for which see: De Antonelli – Guzzi – Ferroni – Slocovich in "*Informatica*" (AICA XVIII 1988–2) pp. 199–222. From this emerged the need for thorough defining specifications, which I will now outline.

1. Database is to Book what Documentary Informatics is to Linguistic Informatics (see my "Foundations of Linguistic Informatics", *Vita e Pensiero* 1987 n. 66–69 and 74–75). Linguistic Informatics aims at inventorying and classifying whatever constitutes a text or discourse, while Documentary Informatics is a catalogue of information where one can find whatever one is interested in. For example, Databases include encyclopaedias, dictionaries, yellow pages, telephone directories, the ancient "*silvae rerum*", the not so ancient "Roget's Thesaurus" etc. Whereas poems, novels, histories, essays, treatises, manuals etc. have always been called "books". Between the two extremes there are obviously gradations with indefinable boundaries. Nonetheless, it would seem that the difference between these two poles might lie in the unity of the whole that is imposed by their different functions. The Database is an agglomeration with a single order that allows one to find information. The book is a single thing: it has a unity of form, a unitary form of the whole, which makes a particular book unique and different from all the others.

The Database is a storehouse of building materials; the book is an edifice with its own peculiar architecture.

2. Database is to book what consulting is to reading. The difference between these two, which are always simply two ways of reading, is obvious, but the algorithms of reading as opposed to speaking are as complex and difficult to list as they are easy to run. A database is only consulted, never read; a book, however, may be both read and consulted. In fact, every book is always a mine of information. Yet in a book there can be pieces of information that need other pieces of information, and it is at this point that hypertext comes into play.

3. The name Hypertext is new, but the idea and the function are not. Concordances, parallel passages, interlinear glosses, marginalia, bilingual and multilingual dictionaries existed before the invention of printing, without disturbing the "international tablets" of commercial values that were used at Ebla in 3,500 B.C. For example, when one reads Dante, one encounters a host of personal names: any information added about these persons will be a hypertext. However, an informatics expert who had memorized both the *Divina Commedia* and its commentators and the *Enciclopedia Dantesca* would use the word hypertext of a program that:

a) having connected the verse "*galeotto fu il libro e chi lo scrisse*" (Galehaut was the book and the person who wrote it) to where it was explained by others,
b) allows anyone immediately to call up the contexts in order to find out whether Dante was or was not referring to a book or author with a definite title etc.

It goes without saying that the connections to be predetermined are chosen on the basis of the probability of the number of requests. In this case, which is the most common, the hypertext therefore adds to the text information that otherwise would not be found there.

4. The hypertext of the *Index Thomisticus* is of a completely different nature. It is still a true hypertext, indeed, for me, it is one whose value is more profound and more essential. In fact, it does not add pieces of information, but makes them explicit through the addition of codes. These pieces of information are there in the text and the reader grasps them by reading, but the computer cannot work on them unless they have been previously formalized.

Those of the *Index Thomisticus* can be reduced to two layers: the typology of the discourse and the lemmatization. The first one specifies, for example, if it is a quotation, and whether it merely quotes the sense or is a word for word quotation, or whether it is the title of someone else's work, or the title of a chapter or of an introduction or of a conclusion etc. The second specifies, for example, what part of what word a particular word is; what other meanings it might have, what case, mood, tense or person it is etc.

Pieces of information of this sort exist in the text and therefore do not need to be derived from outside it. But their existence there is comprehensible only to a human reader, it is not also operable by a computer, given that only letters and punctuation have been loaded into a computer's memory (seen. 72–73 of my manual mentioned above). When one reads, one understands that in the sentence "dogs must be kept on a lead" the word "lead" means one thing and that in the sentence "lead is a soft metal" it means something different, but the computer can only distinguish between the meanings if it finds them in its memory already distinguished by something else.

5. The capacity and durability of the CD-ROM has been splendidly emphasized. Its bits, in fact, are not magnetic, but physical hollows produced by a laser beam on a veil of metal. Moreover, everyone greatly appreciated the facility and rapidity with which the computer answered the requests, improvised, not prepared, of the audience. For example, it informed us in a few seconds that the word "*textus*" is used 16 times by St Thomas and the word "*textu*" 19 times, while the word "*textibus*" never occurs in his writings; neither "*textus*" nor "*textu*" is ever used to mean "textile", they are never the participle or supine of "*texo -ere*", but always from "*textus -us*", and so on. Obviously only the experts in the audience knew how the logic of all this had been delegated to the "machine" by the authors of the program both for installation of the text and hypertext in the CD-ROM, and for research into and recall from it. And of course, there is always someone who, in wonder at the novelty and the unexpected, ends up by attributing the result to the machine's own magic, as if it were the machine itself that produces the result and not the people who enable it to do so.

6. However, nobody claimed that the CD-ROM replaced the book, nor did anyone assert that consulting the CD-ROM exonerated him from reading. It was clear to everyone that by their very nature the volumes of the *Index Thomisticus* and the texts on CD-ROM are complementary. Yet I cannot help fearing that the CD-ROM might, almost pathologically, be taken as a substitute for reading both the texts and the concordances and the lexicological tables. The reason for this is that to consult the computer is much easier and swifter than reading and understanding. This is so today more so than yesterday, because today more so than yesterday the pressure is to do everything in haste, and this inevitably leads to superficiality. We all complain that information suffers from inflation. Certainly, as always, the very nature of things in the end will bring about an equilibrium even between electronic books and printed books. But it would be wise to smooth the ways of wisdom, by repeating and emphasizing didactically, especially for the young, the following points.

7. In a book, as in a picture or a concert, the human spirit aims especially at the immediate apprehension of the formulas of the whole. From these it descends to the analysis of the details, and thence re-ascends to a more profound understanding of how those details mingle together in the whole. The Database of a computer helps us precisely to define individual details one after another. Moreover, in the sphere of Linguistic Informatics, it first of all processes—so to speak—each and every individual one of the verbal micro-elements of the text, and then it

sends them off, with their classifications, to reassemble themselves. … But the intuitive understanding of the formula of the whole can only arise, however it may be and, in every case, from reading and re-reading. Perhaps this understanding will turn out to be more profound and documented if prepared and preceded by computerized analyses, but it can never be replaced by them. Without the study of anatomy medical science could not exist, but medical science is not merely the study of anatomy.

8. Furthermore, and lastly, implicit in reading is the exercise of hermeneutics or interpretation according to two coordinates: one is the global meanings of the text, beginning with those of the individual sentences, the other the meanings of the lexicon employed therein. Interpreting involves comparing our sign-concept connections with the sign-concept connections that existed in the mind of the author. To verify the correspondence of the individual words the only possible resource is to examine the part that each word plays in the whole, i.e. in the context, the sense of which is "grasped" by reading. The interpretation of contemporary texts and of the same language and subject area can be easy and instinctive. This is not the case when one deals with text published seven centuries ago, such as those of St Thomas. Seven hundred years of historical and cultural evolution have not taken place for nothing: the lexicon has evolved just as the points of concentration of mental interests have.

I repeat that every valid interpretation unravels in four phases: reading of the whole; analysis of the details; their classification; and re-reading of the whole. Only thus was it realized, for example, that for St Thomas the expression "*ratio seminalis*" meant what we mean by "genetic programming". Yet had I wanted to look for the concept in St Thomas, what word should I have had to call up from the CD-ROM? It would immediately have told me that the words "*programma–atis*" and "*geneticus–a–um*" do not exist in St Thomas...

A very bad use of the CD-ROM of St Thomas, which I deprecate and fear, is that of seeking Italian words with Latinized endings which correspond to one's concepts of today, without first having tested their semantic adequacy. Indeed, even worse, it has already happened that someone has attributed to the personal doctrine of St Thomas the very objections that he reports others have made to his very own doctrine …. Dante wrote of "... *la fretta, /che l'onestade a ogn'atto dismaga* ..." (*Purg*. III, 10–11) (...haste, which diminishes the dignity of every act...). The speed of the computer is definitely not meant to favour superficiality, but rather accuracy, reflection, and methodicalness, no matter how long it takes.

Chapter 15 To Do and to Cause to Do: Man and Machine

Roberto Busa S.J.

Busa, R. 1991. Fare e Far Fare: Uomo e Macchina. In *Intelligenza naturale e Intelligenza artificiale – XLIII Conv. St. Fil. di Gallarate – Apr. 1988,* 57–77. Genova: Marietti.

Editors' note
This article sees Busa return to a familiar mode of argument about the capacity of technology to confront us with gaps in our understanding of what it means to be human. Here he argues that Artificial Intelligence (AI) research has brought to our attention how little it is that we understand about Natural Intelligence (NI). Accordingly, and seeking to contribute to the Philosophy of Science, Busa discusses a number of concepts that are fundamental to NI and by extension AI, including the nature of activity and the author / work relationship. He proposes that the question of the kind of relationship that exists between NI and AI can be answered by undertaking a study based on the *Index Thomisticus* that "essentializ[es] the Thomistic lexicon of acting and of doing".

1 To do and to cause to do: cause and system

In my opinion, if one were to compare the two intelligences, the natural and the artificial (henceforth NI–AI), one would find two methodological lines, which I consider complementary because they start from opposite positions. The first, and I confine my remarks to this, is to collect data that are generally held certain, a certainty that I call generative ontology, and then to apply them to the binomial, "thought-machine". The second is to collect instances internal to AI to find therein in concave that which philosophy, or rather, the reflex ontology of action affirms in convex, as does, for example, the Rev. Prof. Basti. The first is a reflection on obvious universal macro-logical certainties. The second seems to be methodologically more scientific in that it is micro-logical. It is also more respectful of other people, yet attainable only by those who are fully acquainted with the lexicon and the discourses of the contemporary philosophy of science.

1.1 Towards a general theory of activity

Some investigations have brought me to think that in contemporary culture the following observations are not usually explicit. Firstly, each "thing" is already a system or structure of "causes", or components: two "internal", the cause material and formal, and two "external", the cause efficient and final. Secondly, as a root of

J. Nyhan and M. Passarotti (eds.), *One Origin of Digital Humanities*,
https://doi.org/10.1007/978-3-030-18313-4_14

this design, all reality, i.e. the whole cosmos, is a system of forces that are (also) linked, united, connected and conjoined to produce each and every transformation of nature, both evolutionary and involutionary. Thirdly, in every organization of productive forces, some forces sustain endurance (the "*causae essendi*") while others set things in motion (the "*causae fiendi*"). Some are forces that use (the "*principales*") others are the forces that are used (the "*instrumentales*"). Some act through the force that is within them (the "*propriae per se*") others through co-involvement, repercussion, or coincidence of acting with others (the "*per aliud / per accidens*"). It would seem that today the sciences mainly develop the list of the interactions of the constituent parts of the corporeal wholes (the "*partes integrantes*"). But I do not know whether they explicitly consider those complexities that St Thomas sums up as "*actus et potentia*". Whether that which everyone says is "*in actu exercito*" were no more than to see to it that every period of one's own speech should be syntactically and paratactically correct, every sentence grammatically correct, and that every word should have its correct inflection.

Moreover, the mental pictures of the kinetic concept of the material structures seem indeed to contain the "cause to move", but not the other universal values of the "cause to be, cause to live, cause to think, cause to love, and cause to rejoice". There are diverse human disciplines that deal with these, but not sufficiently widespread is a theory of the activity, i.e. of being and causing to be, which is general and ontological. If anyone were to undertake the long and arduous labour of essentializing the Thomistic lexicon of acting and of doing, using the documentation provided by the *Index Thomisticus* as a guide, as well as a magnificent service, he would also thereby provide both a treatise on the epistemology of the action of every being, i.e. of "*onto-logy*", and an analysis of the language. There one would also find the answer to the questions posed by the relationship between NI and AI.

1.2 Activity in general

Every action is a structure: a network of presences between more actions, actual or possible. Every being implies activity. Every being is an active presence. Every activity is always a bipolar arch, a relationship between one thing and another, a bridge between two banks, the single straight line that connects two points. But we know, both from the natural sciences and from the revelation of the mystery of God, that the compresent and interactive multiplicities are never just two, but innumerable. And at the very bottom of everything the Nature of Being is Three Persons.

However, in our world an enormous mass of micro-elements and micro-events has coagulated and clustered into a gigantic number of systems which then at the very end are just one. Each and every one is inserted into the sole universal cosmic system, each attached to the other with the infinite varieties of those moduli, heteroclite in themselves, that are the "four causes". In the universe everything is interactive and interdependent with everything else: some mediately, others immediately.

1.3 Activity efficient

I will confine my remarks to the activity that is productive or efficient, that which brings something else into being. The subject matter constrains me to forego any analysis of those of our activities that are the fruits of our joy, "being brought into being" by another, with love, admiration, and appreciation of beauty. Every act consists in producing in someone else something that the author already had. This holds for movement, heating, forming, informing; it implies the multiplication of specific unities.

I will now analyse the elementary formula. In A there is something ("form") that is lacking in B: i.e. A has such a value "*in actu*"; B merely "*in potentia*", i.e. in receptive possibility. "A" with that shiver which we call "activity" (which is an "*accidens*", a moment, a frequency, a wave … but not something that is emitted or a leakage, but rather a completion, a sign and fruit of abundance) "duplicates" this form, i.e. it causes it to exist also in B, in which that form is the reflected expansion of that in A.

Now, in A this "form" (whatever it be, whether quality or value) is either a part of the nature of A, of its very consistency, from which it originates as a property, or is detached from it and different. If it is different, it thereby suggests that it has arrived there from outside, or that something else has put it there, i.e. it is produced or caused. In this latter case, A is merely an intermediary transmitter. But in the first case, A is the source, the real cause, the intrinsic cause of what it communicates to something else. Thus "*compositum habet causam*": that which in a subject does not derive from its own nature, i.e. does not come from within, but has come from outside, or from something else.

Making, i.e. causing to exist, is therefore, existential, factual, actual: presence in existence. This can be seen in the Thomistic expression that is so difficult to translate, "*in actu*". To cause to exist is not the same as creating an idea, a possibility. The possibilities of other persons and other things are "scissored" (parti-cipations) by the First Fullness of Everything, which of itself and with itself alone brings into being other things and other persons, with a freedom creative of clippings and cuttings, combinations and associations, and compositions and assembling, unpredictable in that they are new, i.e. not deducible from those that already exist.

For the same reasons all logics, even the formal ones, prescind from actuality and are contained in the world of what is possible and what is thinkable. Just like mathematical calculation, every other logic too in fact separates and prescinds from dynamic and causal relations, in order to occupy itself solely with relations between the whole and the parts and between parts and other parts. It is the doing that brings about a comparison between two beings, which are opposite and connected like author and work, or at least like activity and the object produced. It is the cause efficient that multiplies: there is a multiplicity of beings, if there is someone who brings others into being.

1.4 Author and work

From the same reason, i.e. that the idea of one nature does not still contain its very own existence (distinct from and correlated with that existence whose nature it possesses as its own) derives the fact that we all know that within it there are causes that are authors and causes that are not authors, but rather, if at all, transmitters, repeaters, copiers, interfaces. In the author the operative arch is twofold: to think of the idea is one thing, to bring it into being is another. First, invention, conception, planning, then, production and construction, i.e. causing to exist.

This is not the place to analyse the mysterious mechanisms of the birth of new ideas. The fact remains that the mind, as well as continuously measuring whatever it encounters, or rather whatever already exists in reality, every now and again finds itself in possession of something that no one–as far as it knows–has previously either thought of or brought into being, i.e. an assemblage of components that indeed were already known, and perhaps by others, but assembled according to a formula that is indeed new and aims at a final result that is also in some true measure, and thus under some aspect, absolutely new. I merely observe that the paternity of a fruit of the intellect is a value that is not only irreversible, i.e. cannot be overturned, and is not reciprocal (on the contrary it is one-directional). It is also everlasting, i.e. indestructible: it is in fact tinged with existence: "*quod factum est, infectum fieri nequit*". Everyone knows this, thanks to patents and copyrights of authors and editors, human activities behind which stands the justice of my money / your money.

In Holy Scripture God too claims His rights as first author, but behind these stands only the law of His love as father. However, a second author is not "*causa essendi*" but only "*causa fiendi*" of his work. First, he has the intuition of a new assemblage of various components in a new finalized unity, then he analyses and designs this idea in all its details. Then he constructs it, finding the raw materials and pieces for it, and lastly, he connects everything with a source of energy. If the inventor stops thinking about it during the assembly, the assembly stops; but if he goes away after finishing it, his work continues to exist "on its own account", although it will always be his work.

Therefore, the work-author relationship is always active between the inventor and even every reproduction of his product. To everyone else the work will appear as something unrelated, but yet one still has to deal with it. I imagine that the myth of the man-machine antagonism arose because there was no inventor seen to be between the machine and the human being, or rather because of the social frictions caused by machines that produce negative effects on people.

1.5 The "system" of the forces

St Thomas expresses this with the words "*causae per se ordinatae*". For him "*ordinatae*" means "organized", hierarchically structured in a finalized operative and co-operative system. "*Per se*" means by force of its own nature, not by lateral and incidental involvement, not "*per accidens*". By "system" I mean a multiplicity of different things put together to achieve that common result, the whole of which is finalized. For example, a poet sits writing at a table. The forces in play are extensive: his imagination with its poetic intelligence, his deliberation as to how to put it to work, his knowledge of the language, the psychomotorial apparatus of his whole organism from the head to the arm, the fingers, pen, nib, ink, paper ... . The "*per accidens*" are the telephone ringing, a fly buzzing, an interior anxiety that suddenly occurs ... who could ever count them all? It is, however, obvious that such "*per accidens*" are the result of the interconnections of the subsets in which the poet is integrated. It is only within whatever system that so many of the modes and types of causes that recur in St Thomas and scholastic texts have any value.

1.6 Activity that is fruitive and activity that is merely transmissive

I give this translation, although not perfect, of the scholastic dictum "*operatio immanens—operatio transiens*". The first enriches anyone who carries it out. When the agent acts in favour of himself, to preserve or to enrich himself, with his own activity he attains his own objective: "*movet se ipsum*", in other words he modifies himself. And thus "*actio est perfectio agentis*", the action is the complement, the perfection of the agent. Every fruitive activity is such, and such is every expressive activity. It may be affirmed that the boundary between living and non-living is precisely here: every operation of the living is always and above all "*immanent*", yet with real influences on others. Every operation of the non-living is merely "transient". However, an activity that works in one's own favour will always be the "principal" of its own system, and will always be part of a more extensive and superior system.

An operation that merely passes is that which, since it does not aim at the acting subject, does not enrich him, neither does it arise there. It comes from outside and thither it returns. Such is the operation of every transmitter or interface or repeater: like a wave of the sea or an electric frequency on a wire. "*Non est perfectio agentis sed ipsius moti*", i.e. it makes what it meets work, but not as if it were the source that starts it off, or as the final result in which it finishes

1.7 Forces that assemble something; forces that keep something in being

I give this translation of the scholastic expression "*causae fiendi*" and "*causae essendi*". "*Causa fiendi*", or that of the beginning, is that of one who assembles things or events, who brings about the exercise of the "*causae essendi*". The builder who puts the ceiling on the internal walls that will continue to support it even when he has gone. When I flick a switch, I cause the electric current to start and to continue the incandescence of the filament of the light bulb. I am the "*causa fiendi*" of the illumination, while the electric current is the "*causa essendi*". A motor car continues to exist independently of the assembly line that put it together. Even such a binomial has no meaning within the abstraction that analyses an elementary relationship of cause and effect; on the contrary, it supposes a system of forces converging on one goal.

1.8 Forces that use and forces that can be used: owners and executors

In this expression is implicit the observation that any instrument, with the exercise of something proper to it, can co-produce by means of someone's force, something in which it has no interest of its own and which transcends its own proper and natural capacities. In the hand of the poet the pen writes poetry. As a consequence, there are in the product as many aspects as there are causes or forces that have collaborated to produce it. In the poet's sonnet there are traces not only of the quality of his creativity but also traces of his hand, his pen, and his ink. I deliberately say "aspects", i.e. "forms", that are qualities, dimensions, figures etc, because I deliberately do not say "parts". The pen of the poet writes the whole of the poetry, as indeed does the hand that holds the pen, the brain that moves the hand, and the mind that moves the whole. So, I will call "principal" that force that of its own and on its own has contributed to bringing into being what is therefore considered principal. It is always the objective that acts as "commander". In literary criticism the commanders will be the imagination of the poet and his knowledge of the language; in calligraphy it will be the quality of his psycho-motorial apparatus, etc.

1.9 To conclude: more ways of causing to do

As well as "doing", there is therefore "causing to do", when a force uses for its own proper aim the exercise of abilities proper to other people or instruments. There is also a way of acting so that somebody else, and not the first initiator is the "responsible owner" of the result. In an examination it is the candidate, not the person who has called him for examination. When one tries out a typewriter in a shop before buying it, it is the machine that is responsible for the quality, not the writer. Upon the implications of every system of organized forces St Thomas bases the "philosophical" certainty of the continuous presence of God in every reality operating in every event. He illustrates it with examples that he takes from macroscopic observations of the human life and the nature that he knew.

Today, Informatics too brings us all to notice that every operative structure needs first to be programmed i.e. that logic is the other end of a source logic, that every organization has first had to be thought out. And that therefore "*omne compositum habet causam*", that every system has had an author.

2.0 Man and machine: thought and computer

2.1 Natural intelligence and artificial intelligence: author and work

I consider it opportune to analyse the man-machine relationship as a relationship of author to work, without, for the moment, having recourse to the most valid notion of "instrument". That is abundantly obvious to everyone, if one but reflects on it. Every piece of technology, too, is a human expression, as is speaking, the arts, behaviour. Between man and machine, according to their very nature, opposition and rivalry, apart from being myths of journalism, are also words without meaning. *Aida* posed no threat to Verdi, nor did the radio to Marconi, nor did the *Divina Commedia* to Dante. No work is ever a rival; it is always a manifestation and a glory. The author-work relationship is operative, it starts from one reality and arrives at another, thought out and produced. Every operative arch is a one-way system. Its two poles are heteroclite, i.e. irreducible and irreversible, as are the terms of every relationship between them (and the relationship does not go from active to passive, but vice versa). The author is first and is the starting point; the work is second and is the point of arrival.

2.2 Invention: interior intuitive intentionality

Author is a synonym of "inventor", that term that is perhaps more appropriate in the field of technology. *Inventor* seems to mean the person in whom the idea of a new whole, of a new machine, first arose. I say "whole" because every machine is a unitary structure that combines different parts, adapted to each other and unified by the completion. I say "arose", not as if its absolute possibility was thought out for the first time on that occasion, but because the idea-formula occurred in that particular person, for the first time in history as far as we can tell.

What can be categorized as an invention and a new formula of organization? I present it according to three values that are connected and interchangeable: interiority, intentionality, intuition. By *interiority* I mean:

a) conscience, reflexivity about oneself to oneself and
b) about oneself as a "database" tending to give, to offer and exchange with others.

By *intentionality* I mean that all conscience is also a possibility "of" someone else. It is *wanting* to be able to dominate, communicate, and reproduce. Intentionality means that my conscience, added to executive will, is an operative principle "of" someone else and "on" someone else. But let it be noted that it is one thing to devise the form, quite another to give existence to that form. No man has ever known how to bring into being another being *ex integro*; he merely knows how to transform what already exists.

I call three operative modules of our conscience "intuition". First is the fact that we experiment, we see, we touch the presence of others and that therefore, the existence of each one of us is co-presence, correlativity. One gets the impression that being/existing and presence are equivalent, are synonymous. Secondly, I call "intuition" that mental dynamism with which, after a still non-existent goal has been formulated, we devise the idea of how to achieve it. Thirdly I use the word "intuition" (or "abstraction"?) for the fact that for any object, we individuate and grasp, making it intelligible and indefinitely multipliable, some aspect of it, i.e. form or quality, that we reformulate and accumulate in our "*intellectus possibilis*".

2.3 Natural intelligence invents i.e. programs artificial intelligence

NI has always been so intelligent as to know how to make use of that, let us say, "*online*" machine which for each person is his body. But it has always been so intelligent as to also construct the "*offline*" extensions of its own body that we call machines and instruments, to do more and better than what could be done just with its own body. I absolutely do not see any scandal or any true novelty in the fact that with his machines man arrived at results that would have been unattainable without them. If not, what is the purpose of technology? It has existed since the beginning of the world. Were clothing, cooking, cultivation, defence and hunting, jewellery and perfume etc., not all in their time *avant garde* technology? The history of Robinson Crusoe is a summary of this.

It should always be remembered that every production of instruments has been preceded by an invention or ideation or programming. Electronic technology can elaborate any piece of data and information that is formalizable, or if you prefer, formalizable and thus representable in bytes. As a result, all programming starts from the micrological analysis of events of which man in the course of his life has apprehensions, and macroscopic operations, thanks to the intuitions of the whole that I mentioned above. He first of all crumbles every event into its minimum elementary components, represented by strings of impulses (or equivalent entities) that are different only in the arrangement of combinations, to which a human being attaches the semantic values of numerical codes. He transforms, i.e. represents, every quality, type, class or category with long trains of numbers in bits, i.e. just in zeros and ones, that are characteristically different. He then proceeds to reunite these trains into wholes on the first level and these into wholes on a higher level and so forth, one elementary operation after another. It is the finalization which, from the innumerable possible combinations of elements in wholes, chooses the few "that serve the purpose".

Informatics obliges all its operators to analyse in this way all knowledge and every logical inference. Our logic, thanks to the machines that she herself produces, challenges herself by electrically renewing the ancient adage "*gnothi seauton*". It may be pleasant to speak in macroscopic and perhaps journalistic terms about such a reflexive effort, but it is hard work to carry it out and perform it in terms that are precise enough for the machine to execute it usefully. Anyone who just talks about it may sometimes perhaps talk nonsense, but someone who does it knows how even the slightest single oversight leads implacably to failure and, when his program finally "goes", knows what it has cost him.

2.4 "Source" artificial intelligence is the same as natural intelligence

All AI always and in any case arises as the exercise of interiority of NI on itself, which, because it is reflexive on itself, is also intuitive of others. It is a macrology that unravels itself and crumbles itself into systems of elementary algorithms, into dust particles of very small micrologies, within global strategies which in the computer "make it produce" recognitions, inventions, choices and decisions. The bipolar arch of the production is in fact preceded by that of the ideation. Ideation starts from the goal that is perceived and desired as achievable; from here it proceeds to show how it may achieve that step by step using the possibilities of the computer, Thus the NI of the programmer transfers to the computer all (both at least potentially entire and complete) the strategy which it has devised. When I say "potentially complete", I refer to those algorithms with which NI causes a machine to produce results which it could not even predetermine individually. Although it does not know how to have them made by its own body alone, it does know how to have them made by its machine. All programmers know well that it is the ideation that is more valuable: it is not for nothing that they demand such high salaries.

Therefore, it is NI that first of all makes itself artificial, in order to be embodied in the machine, to invade it and permeate it, as an executive instrument of its own logic, i.e. of itself, since it has logic as its very own force, light, and form". AI, that of the machine, is always an expression and thus a proxy of the micro-logical interior self-reflection of NI. Between humanism and technology there is not so much a struggle of class as of integration. In my opinion it is obvious and fundamental that NI and AI should not be considered as alternative, as if they are not the source program and target program. They are the points of departure and arrival of the one same operational logic. NI and AI are—and I cannot say this often enough—the two poles of the same operative arch, in which the second pole is inseparable from the first. It is well established that every inventor or author or programmer is merely an assembler, one who for the first time puts together ideas and notions, and then things and parts, which then continue to exist and consist independently of him who is their "*causa fiendi*" not their "*causa essendi*".

2.5 What a human being cannot delegate to a machine

It will be noted that I insist on the verb "delegate". The NI–AI problem is out of focus if I contrapose a computer to all those who have never made one, or a program to all those who have never written one. Every machine must be matched with the person that conceived it, programmed it, or built it. When I see a chessboard, I do not say to myself, "does one win or lose with this?" As an instrument of logic every computer knows and acts according to how much logic has been installed in it by its inventor, builder, or programmer. Indeed, these people are only assemblers or directors ("*causae fiendi*"), but each victory must be attributed to the players, not to the spectators...

I insist on saying that the author-machine operative arch is inseparable. Therefore, I am using language perfectly correctly when I say "a computer can be programmed to program and to discover and to choose". However, instead of saying "the machine cannot think" one should say "no human being would ever succeed in making a machine that could think". To say "*man with machines* achieves results, *the magnitude* of which without them he would not be able even *to predict*" is more precise than to say "the machine obtains results unpredictable even by man". The sentence "the machine educates man to think algorithmically" would be better put as "with the machine man educates himself to think algorithmically". In other words, to explain the matter in terms of man-machine opposition would not be correct. What would be correct, however, is every sentence in which it is conveyed that the machine was composed in a system with its human inventor.

Then I avoid saying "something that man cannot do", but I do say "something that man does not succeed" in delegating to the machine, so that it is clear that it is a matter not of illegality or desecration, but rather one of impossibility, i.e. unthinkability and illogicality. In view of the above, I wish it, in fact, to be understood as self-evident that thought does not succeed in delegating to any machine the fact of being its own first author. I may perhaps cause one of my machines to invent something else, but it will always be me who exercises the author's rights over what the machine devises. Obviously with this assertion I comprehend all that which it implies, and which I do not list here.

By definition, the fact of being a "first" cannot be delegated. A delegate is always a second. In a railway train there will always perforce be a first carriage and a last. So, an "intentional form" (or fact or value or activity or thought etc.) is always a first (though not in absolute) principle and point of departure. It is in fact an activity in itself, because it is a reflexive awareness indeed that it itself exists, but it is also open and ready to operate on other possible values of existing in others. In this sense it is a one that is a principle of two, i.e. operating on a two: it knows itself plus another.

Even the word "I" signifies something incommunicable. It is in fact a deictic word, and signifies awareness not of a concept or nature but of a presence. Only I can say the word "I" as I say it. Every "I" in the mouth of someone else is for me a "you" or a "he", i.e. the "I" of someone else, but not me. So, the intentional is a whole composed of me and someone else. Of this I could indeed delegate half, the other, but not the "me" as well. The root of the matter is that "first" and "I" are a tautology and are such at least every single time we reflect on communications between persons, incommunicable singularities, in the sense of being singularly unrepeatable i.e., indeed they are many and distinct, yet each one cannot be multiplied. And they communicate and pass on to each other forms and values of life and being.

In less general and less synthetic but more current terms, this same impossibility is repeated in the binomial "author-work". A sentence such as the following would have no sense at all: "a human being programs a machine which first programs itself and of itself decides to operate." Such a machine would not be a computer but a programmer. And if we do not attempt to produce human beings artificially, it is not because we want to curb the population explosion, but because such words have as much sense as if one were to say "Verdi would have done better to compose operas that did not have a composer". To be the author of something is a great attribute, both sacred and mysterious. It is in fact the sign of a presence and participation of the Infinite, which is, let us say, special and superior. In fact, my definition of "infinite" goes as follows: infinite is he who with himself alone knows how to bring into being *ex integro* an "other" who is not him. An absolute "I" who knows how to bring into being another "I" who is "I" only for himself, for whom he himself will be a "you" and a "he" and vice versa. Who could ever stop such a one? And who could doubt that there are so many to be "I"? In short, a work can indeed be transmitted, reproduced, copied, i.e. recreated by others, but he who finds or receives and thus copies or reproduces it, does not become its author. Between origin and transmission, between source and interface, there is a well-marked distinction.

In current culture the awareness of the peculiarity of the author-work relationship is lively, indeed most lively, as being the source of the author's rights. It is not so in scientific culture. This seems to concern itself only with energy in transmission or transformation. Here the notion of first origin or first source of energy is not often to be found. Thermodynamics, at least as I understood it fifty years ago, had as a second principle something that is impossible to realize, a *perpetuum mobile* of the first kind, i.e. which produces work without removing heat from any source. A *perpetuum mobile* of the second kind, which produces work in a closed cycle and takes heat from a single source, is simply impossible. Therefore, physicists are astounded when someone asks them what the primary source or sources of cosmic energy or energies are.

2.6 What a human being can delegate to a machine

To assert that it is possible to delegate to a machine every strategy of operative logic that is formalizable and that can be represented as bytes i.e. expressed as an algorithm, is to say everything and nothing. But is there any knowledge, inferences and values that cannot be formalized? If someone were to say "yes", thinking, for example, of the intuition of the whole, sense of beauty, joy, and loving, I would rephrase the question thus: could those values whose own content was not directly either quantifiable or formalizable be formalized at least by concomitants and consequences? For example, could not the joy of an audience after a concert be accurately measured by the volume and duration of the applause? I remember my wonder many years ago when I realized that Trinitarian theology—a splendid monument to severity of method and greatness of content—was summed up in simple arithmetic: "*una natura, duo processiones, tres personae, quattuor relationes, quinque notiones...*". Even today I cannot forget that the Logos became flesh...

I believe that the satisfaction the Lord will have when He notices how skilful His children are in AI will compensate Him for the materialistic superficialities that some have cast in His face. The whole problem is much more to do with the anthropology of NI—it shows us how little we understand it—than with the technology of AI. Machines are made of bodies and events, not of hypotheses and still less of words. Since there is no harm in trying, it will always be useful to set the most daring research objectives. Just to speak of them merely for the amusement of science fiction, the new literary genre of the electronic fairy tale. Therefore, I will say after Monsieur de la Palice that everything that a machine can be made to do can be delegated to it.

Therefore, let AI do as much as it can! But if anyone were to tell me that Dante Alighieri was invented by the Divina Commedia, it would not be anything to do with technology or AI, but simply some schoolboy nonsense of the type: "I was alone—with 100 others—seated and standing on a wooden stone—when a boneless skeleton—with a hand that it didn't have—struck me with a knife that had no blade or handle...". These are words that can be written and said, but not things that can exist in reality. Thus, terms that initially seem to be obviously contradictory make research into them utterly irrational. But even if it is only at the end of the research that one realizes that one has been dealing with a real contradiction, this nonetheless would make the research useful and worthwhile.

2.7 Are there limits and barriers?

It is true that we, merely by speaking and writing in some measure about everything, imply that everything in some true sense can be formalized. Language is the first and natural formalized expressions of signs. The "sign" (in itself bound not to knowledge but to its communication) is a paradoxical reality that is pregnant with meaning. It has enormous force and dynamicity. Speaker A addressing the receiver B with a sign C communicates a piece of information D. Of these four ABCD, none is any one of the others but it is with this algorithm that two people speak and understand each other.

Yet actually linguistic informatics at least is, in my opinion, marking time in front of a barrier that it can scarcely get over, except with much toil: the formalization of the global meaning of a whole. In every text the global significance of whole, the whole of everything is trabeated in its turn by so many sub-wholes, passage by passage, sentence by sentence, clause by clause. Every whole of meaning is undoubtedly represented by the whole of its meanings, i.e. by all the lines and all the words. But can some few signs or distributions or something else, I know not what, be found or inserted there that are characteristic of the whole, just as one can deduce from the pulse the physical condition of the whole organism?

Our mind, somewhat confusingly, intuits first the whole and from the whole descends to determine its single components. With the computer, however, I ascend from the elementary dust particles of the smallest components towards the whole of everything. The semantics of the whole, however, is more than the sum of the values of the individual components. Now it would be enormously useful to be able to construct algorithms on data that specifically represent the semantic whole, but are much smaller that the whole text.

I do not know whether formal and algebraic logic have identified the factors that allow the formalization of the whole: which ones exist only within certain limits of proximity and which exist at any level of comprehensiveness.

3.0 Thought and brain

3.1 The problem; facts and interpretation

It might be thought that informatics and electro-neurophysiology present opposite tendencies. Indeed, informatics documents the priority and power of logic over structure, system, organization and number: logic is their source and their master; thought dominates the material and expresses itself with it. Electro-neurophysiology, on the other hand, sometimes seems to suggest that it is the physic-chemical material that dominates thought and conditions it. In fact, many scientific experiments show that determinate modifications, renewable in the laboratory, of the electrical and chemical states of nerve structures alter or excite or extinguish determinate human and animal physical behaviours. The very word "psycho-pharmaceutical" confirms this.

Why is the psyche "commanded" by chemical products? Is it perhaps because it is nothing but chemical, i.e. it has no other source, or rather it has no cause of its own and in itself other than chemical structure? Or on the contrary, why in its exercise is it conditioned by instrumental intermediate forces? If a painter painted badly, how could I infer from that that he was without talent, instead of inferring that he used damaged brushes and poor paint?

Experimentation and its description are usually left to neuroscience, but philosophy is required to provide its interpretation. This is obvious and true: provided we do not forget that "philosophy", at least in the sense of "ontology" is already present and operating within any research and any description of it: in fact, it is present and working there, i.e. implicit, as a weft or transporting tape: as logic generative of action and expression. If it were subsequently defined thoughtfully and knowledgeably, and thus expressed correctly in deliberate cultural forms, we should have the philosophical interpretation of these facts, in the current meaning of the word.

The problem of the spirit-matter relationship is very deep and difficult: every human being, with his heart and his life, is a piece of data to interpret. Body and brain are really a "machine", live and *online*, of thought, soul and life. Our consciousness indeed moves our body, macroscopically and within certain limits. But the body is nonetheless microscopically programmed under an enormous number of other aspects by "operative systems", natural software, which elude all our attempts to understand them and remain outside our whole field of awareness and zone of experience. Who will ever be able to define the play of the interfaces between the physical, chemical, biological, psychical, logical, environmental levels that make up the whole of a human being? The functions of our logic are more like those of a periscope in a submarine than those of the dashboard in a motorcar.

In us, spirit and thought are already "delegating" to our gait, face, throat, eyes, and hands. In fact, through our thought our body is already, so to speak, a computer *online*: but it is not given to our logic to peep into our room of our *software*, except with any number of obstacles and still with very poor results. Paradoxically, we know much less about the operative algorithms of our brain, wherein let us imagine our consciousness dwells, than we do about those of our digestion and respiration. On the other hand, every machine is a continuation and *off-line* extension of our body, programmed by our conscious and deliberating logic.

I add just three methodological notes. First, every single event must be studied within the whole, i.e. within the network of which it is the node of confluence of forces. Second, the toil of centuries of Mediterranean Christian culture cannot reasonably be ignored, even though it is obviously merely based on macroscopic observations. The third is a corollary of the two preceding: linguistic informatics today guarantees the methodological value of the electronic micrologic to scientific enquiry into the macroscopic (perhaps it would be better to say maxi-scopic) fact of our speaking and writing.

3.2 Every single event must be interpreted within the whole of the system

The micro-enquiries of electro-neurophysiology furnish most impressive data. These are very small elementary events and samplings defined by complex and laborious procedures that can be reproduced in the laboratory and are thus verifiable. Yet it has always been known that sleep, tiredness and shocks disrupt the course of our conscious attention and command of ourselves. The interpretation of these facts will be strictly scientific only when it is founded on the scientific definition of the whole system of forces of which that event is the result. What should we say to someone who asserted that every movement of the index finger by 3 mm kills a man, because he has seen it done once when a pistol was fired.

Before describing the entire behaviour of a vital system from the fact that an electrode causes alterations in a neuron, logically speaking, all the other factors of which that behaviour is the result must first also be examined, and all the relationships of condition and conditioned must be checked. The enquiry into the micro-element must be integrated within that into the "macro" of the whole. The sciences of corporeal, physical, chemical and biological nature today descend to rummaging in the depths of micro-elements. It was electronics that opened a few decades ago the doors first to the "micro" atomic and electronic, thence to proceed to the "maxi-macro" spatial.

Nonetheless the logic of the relationships between all and the parts, between the whole and its components, means that one must also continue to weigh scientifically the global and comprehensive behaviours of the whole, if only for their macroscopic nature and definition. No structure is simply the addition of its minimum elements. Certainly, from certain aspects, it is a sack of corn. But in a machine, a building, an orchestra, a poem or a painting, the formula, i.e. the form or idea, according to which the components make up the whole, it is that little extra something that from aggregates makes a specific: "(operative) unity of many". Experiments on the brain are still rare lights in an ocean of mysteries. Certainly, they mean something and they certainly must be intensified in all directions: but they still do not justify conclusions that only an overall vision of all the operative structures in play and of all their complex games could support. Finally, if a single behaviour is an object of research that confirms it as a piece of factual data if there are pieces of factual data that are investigated, even the very fact of research, of the investigating behaviour, must also be taken into consideration.

For example, under the microscope lies the field and object of research, but above this there is, on the one hand, the person who researches, who carries out a portion of the strategy that has been thought out and promoted by the director, and on the other hand, there is the office of the firm that has designed, built and put on sale the microscope. There is then also the publication of the research: its description, both verbal and graphic, implies that it is understood that what is expressed there will have correctly included etc. The aforementioned event exists within a system. But each piece of knowledge and every communication of knowledge, whether organized or exploratory, whether classificatory or descriptive, is a system that exists. Being and knowing are not opposite like to be and not to be, but like humans who are known and humans whom one knows, some beings and other beings.

3.3 What has been acquired already must not be ignored, even though it has been acquired only macroscopically

I prefix an epistemological note. Conclusions that are validly drawn from micrological analyses (i.e. computerized ones) will never contradict conclusions validly derived from valid macrological analyses of the same whole. In fact, the same logic of being lies at the base, i.e. it gives origin and life and validity to the macro- as well as to the micro-analyses: the critical and controlling logic that causes them is one, the logic of being. Every man is a man because he possesses it through nature, not through information, and in each man, it is variable, yet one, just like the being.

And in this way, I underline the opportunity that the scientific research and the informatics of AI offer us today. This an opportunity to restore cultural vitality to all that two thousand years of Christian philosophy has acquired and tested precisely on the human soul, with a most honest severity of philosophical method, through generations of scholars, united as a team, in the flow of culture, by the commitment to live its conclusion for themselves and for others. Such precision sums itself up in the statement that the human soul, which succeeds in moving in the discontinuous awareness of itself, is the immortal substantiality and personality of each individual and the "only substantial form" of its own body. Only God knows how to create it, then to call it supernaturally to participate for all eternity in His Life. Such conclusions have been so much more seriously scrutinized with understanding of love, inasmuch as they are connected with the theological implications, for example, of the Incarnation of the Word, which has in personal unity assumed a body and soul like ours, and of the immaculate conception of His Mother. These laborious conquests of the Christian "*psychologia rationalis*" in such a difficult subject, started just from the macrological observation—electronics had not yet been invented! —of the individual and social behaviour of human beings. That is, of all animals, only mankind cooks his food, changes technologies, conducts business, has theatres, libraries, newspapers, has religion, sciences, arts and fashion, seeks security and guarantees, creates for himself methods of transport, is never content with what he has, but lives with dreams of the beautiful and of the whole, of the certain and everlasting. Every living thing wants to live, but only mankind wants to live in a beautiful way, to live on that which is beautiful and true and just and good ... forever. Furthermore, with theological epistemology, the conclusions and the terminology of Christian philosophy (it is "ontology" that makes them communicate) were incorporated in documents of Church teaching.

For example, the Council of Vienne in 1311–12 (v. Denzinger 481–902) resolved that: *quisquis deinceps asserere (...) praesumpserit, quod anima rationalis intellectiva non sit forma corporis humani per se et essentialiter, tamquam haereticus sit censendus*" (from now on, anyone who shall take it upon himself to assert that the rational intellective soul is not the form of the human body in itself and in essence is to be accounted as a heretic). Six centuries later Pius XII in his encyclical *Humani Generis* (12 August 1950: v. Denzinger 2327/3896), recognizing that it was possible to consider that the human body might be the result of evolution, nonetheless affirmed "*animas enim a Deo immediate creatas catholica fides nos retinere jubet*" (the catholic faith commands us to keep hold of the souls created by God without intermediary).

Even for those who have no faith in the Church, such pieces of information still preserve the value of what has been acquired by research that has been scientific and philosophical, collaborative, tested and polished by generations of scholars throughout the centuries. It would not be logical to deny their value *a priori* simply because they belong to the past: even if some cleric were childishly to consider them no more than curios. It is much more sensible to consider them as complementary to modern discoveries of neuroscience, to the mutual enrichment of both. In fact, before anyone declares that the Christian "*psychologia rationalis*" is no longer valid today, honesty demands that it be known and evaluated: and this is above all why it must be translated.

I speak of translation even within the same language. In fact, culture continuously evolves not only from one language to another, but also, even within the same language, in the semantic relationships of word and concept and, perhaps even more swiftly, in the undulation and shifting of the centres and nodes of diffuse mental interests.

3.4 The micro-logic of linguistic informatics

Electro-neurophysiology is a particular, i.e. partial, approach. It must be integrated with the global enquiry into the whole. Therefore, it must be placed beside the sectorial enquiries into other components of the system. Amongst these, language has always had a prevalent and principal position, although not a unique one. The study of human speech until today has obviously been macroscopic and in the nature of things based only on intelligent samplings. It fell to me to realize that computerization would make it possible to apply to linguistic enquiry the methodologies of research such as—although not entirely—those of physics and mathematics. In fact, the computer allows one to elevate micro-logical surveys of countless single elementary particles of discourse to the documental value of their complete census: and that from enormous quantities and masses of texts, written or spoken.

I called such an approach "linguistic analysis" although I knew that this expression has other values in other areas. It is certainly micrological-informatical and physical-mathematical regarding the census of the signifying signs: but it is reflexive and interior, and so "humanistic" regarding the semantic valuation of words, in the whole and in detail. In fact, words turn out to be not at all homogeneous like numbers, but rather strongly heterogeneous. It is certain that language, being an expression of consciousness, is conditioned by the brain, but not solely by it: each act of speaking is the *output* of a system of which the brain is just one of the organs.

To ascend inductively, and, so to speak, in reverse, from the quantitative and the evaluative analysis of speech (and writing and understanding) to the whole system of forces that brings it into existence—without any aprioristic exclusion of what apart from the brain might be part of it—should integrate those enquiries which, in an opposite sense, experiment with physico-chemical modifications of the brain, with a view to measuring consequent modifications of the phenomenon of speech.

Chapter 16 Interior Algorithms of Understanding by Reading

Roberto Busa S.J.

First published as: Busa, R. 1994. Algoritmi interiori del capire leggendo. In *Macchine per leggere. Atti [...] Certosa del Galluzzo 19.11.93*, ed Claudio Leonardi, Marcello Morelli and Francesco Santi, 69–75. Spoleto: Centro It. Studi Alto Medioevo.

Editors' note

In this article Busa explores how one might develop a computer that can read or communicate ideas and understand as humans can. His assumption is that a computer would need to be able to mimic the processes that humans use for this. Accordingly, he reflects on the processes involved in the written communication of ideas from author to reader and briefly on the processes involved in reading and, more generally, comprehending a message that is communicated through image or sound. From this, he claims that he has shown that "the internal process of understanding by reading is the corresponding mirror image of expressing oneself in writing". Regarding the communication of ideas, he describes this process as a unidirectional one from writer to reader. The reader's role is portrayed as being quite passive in terms of the apprehension of meaning—the main thing they do is query an internal table of words and their meanings. Busa supposes that communication works because human beings have been provided with an "identical (or almost identical) conversion table or connection or compatibility between sign and concept". He further opines that computers could be provided with such a conversion table. Yet, he writes that this would be very difficult for various reasons, not least because the computer works via symbol manipulation and so the sign-sense look-up table that he imagines to exist in humans would be a sign-sign table for the computer. It would also be difficult because, as he argued elsewhere, our understanding of syntax and other linguistic phenomena remains quite underdeveloped.

Introduction

I will speak only about communication between people, for in general every action-reaction between objects or nodes is also communication. And I will speak only about communication of human ideas—this is not the only expression, but it is considered the most impressive—and about written communication, not about oral discourse, nor about that which is one-directional, for example, a lesson, nor about that which involves dialogue face to face or on the telephone.

In written communication the two directions, A to B and B to A, are disconnected activities, or rather reciprocal only through the mediation of other things, other times, other spaces: no signal bounces back immediately and directly from the reader to the writer while he writes. In written communication, the surrounding conditions of the writer do not come through to the reader, unless the latter can in-

J. Nyhan and M. Passarotti (eds.), *One Origin of Digital Humanities*,
https://doi.org/10.1007/978-3-030-18313-4_15

terpret some of them from what he reads or from what he knew already. Within written communication I refer only to the reader, and to silent reading, not reading aloud, which is done so that others may listen.

Premises

Communication has paradoxical aspects: it is an operative arch whose two poles are asymmetrical. On the one hand, it is the expression of ideas in words, and on the other it is reading and understanding. These two poles are not like the feet of a stone statue; they are rather the points of departure and arrival of a movement. In play are four distinct and different entities: A who expresses his ideas, B who reads them, C the sign, i.e. what is written, and D the sense, i.e. the concept, message, thought or meaning, that is transferred from A to B. None of the four is any one of the other three. But with the four the transparency between A and B is put into effect. Transparency is an intermediary or rather interface that, in some sense, does not incorporate something within itself to pass it on as a mirror would do. It only transmits something through itself … like light, which, in some sense, tries to let something else be seen and not itself.

Such an admittedly paradoxical algorithm presupposes a fact: that the binomial (the … diode) C and D or rather sign-meaning exists—or at least is centrally identical—in both A and B, i.e. writer and reader, who both know the same language. This in its turn implies that a given set of correspondences of signs-sense could have been acquired by virtue of some natural pairs of signs-sense (signs before words) that exist and operate in every human being that comes into the world. The vitality of such a set of sign-meaning pairs produced the variations and evolutionary complexities of linguistic phenomena within the history of the human race.

Faced with the external message both writer and reader are equal. According to the need to be functional, communication between persons is propositional, not physically imposed. The physical sign, it is true, may demand to be seen and heard, and may even send shivers down one's spine. But the difference between "I am told that" and "I know that" remains irresistible. Nobody can influence the consciousness of another and manipulate it for his own pleasure, except by fraudulent abuse of weakness or psycho-mental distortions. Only the force of the absolute verities of existence have the right and power to bend the consciousness of a human being, which, moreover, is not to bend but to vitalize. And more especially, we all know it is not enough to be right to succeed in having this recognized ….

Of the pieces of information that are communicated, some are (also) operative, such as rules about what to do and what not to do, instructions or advice. Others are (only) cognitive: to know that, to get someone to know that…. These in their turn are subdivided into two fundamental groups: those that are (only) external and those that are (also) internal. The former are completely lacking in the person that receives them; the latter are those values that each person already exercises:

external persons and books do not supply one's first knowledge of these, but they do help to make one realize that one already possesses them.

Pieces of information that are completely lacking are, for example, the historical sciences and the "scientific" sciences, and technologies. The certainties that everyone exercises by living are, for example, existing, living, time, loving one another, the difference between yes and no, between giving and receiving, between doing something and having it done, the question "why?". These constitute the capacity for understanding, the logic that everyone demands be obeyed, the starting of the motor of the dynamism of thinking and of wishing. I am in the habit of calling these "generative ontology", i.e. that which in every human being generates all behaviour that is specifically human. As mentioned above, there are certainties before words. I would describe them metaphorically as the logic incorporated in the hardware together with the operative system installed by nature in that living logical machine that is the spirit of every human being. Mediaeval Scholasticism called them "*obiectum formale mentis*": that by virtue of which thought thinks and the heart of man wishes. Anyone who did not have them would not even be able to understand them except by associationism. Philosophy is the discipline that deals with redefining them as a group, i.e. socially, by placing at the centre of the fire an attention that is analytical, decided and educated, living and being and loving and the same activity of philosophizing that each and every one has.

Understanding by Reading

Within the processes of reading, physicists, biologists, and psychologists are still discovering the various levels and planes of the physic-psychological mechanism that are in play, from visual perception to the recognition of the forms and to the identification of their meaning, both singly and syntactically within the whole. Such analyses have been refined with the techniques of deciphering unknown scripts and with their corresponding cryptographical techniques. Reading is thus a phenomenon that involves the whole realm of nature, and it is an extremely complicated mechanism that appears the more tremendously organized the more one delves into its microcosmic elementary levels, from photons to the atoms of the cells of the retina and of the optic nerves, right down to the tangles of connections within the skein of the bio-psychic forces of the inner senses.

At the end of last August, at the XIXth International Conference of Philosophy in Moscow, I was asked to organize a Round Table with the title "How to do philosophy with the help of computers". My reply was as follows. Such a problem was not new and it already had an answer. In fact, on the one hand, the computer is an artefact with which man prolongs the activities of his hands and his eyes; it is almost a new organ of his body, but "offline", i.e. detached. Its logic, which came from his mind, man has in fact superimposed on materials and external energies assembled by him for his own use, but which exist on their own account.

And on the other hand, all the organic equipment of phonation and audition and writing and reading are simply machines and "online" computers, but internal and vital, nourished indeed by absorbing external materials, but subsumed in the unity of the whole, and operating with vital internal forces that are designed and installed according to an original logic that is not human. I therefore concluded that our title was tantamount to saying: "how can one do philosophy with the help of words, texts and documents". In other words, it was equivalent to asking "do words and books help understanding?".

All this comes back to the actual problem "Mind and body", with the synonymous binomials of thought and reality, soul and body, spirit and material. Thus, the reality develops on three planes or strata: things, knowledge, speech. Of the three (thought, things, words) which is the strongest? Thinking creates new things with technology, assembling materials that already exist; with speech and writing it assembles those other things that are words, which are signs of our awareness of and thinking about things.

As in every other dynamic system, so in the structures of the system of communications between human beings there exists a first organ, whose ignition sets all the rest in motion. I should like to be able to call this very "understanding", which Aristotle had already defined as "that which in us knows how to do everything" (*Peri psyche* III, 5, 430b10ss), and which Scholasticism called "*intellectus agens*" (*Index Thomisticus* v. 21 pp. 1194–1203), the "starter" or "distributor" of the man-machine. Such a motor is not external, but integrated like an organ in the systematic operative whole of the person: the motor is only completed in the whole. St Thomas often commented on a passage of Aristotle (*Metaph.* I, 1 981b7), which he translated as "*signum scientis est posse docere*": one shows that one has understood something if one can communicate it, or rather, thought matures only with the effort of expressing it.

With this I merely wish to point out how human expression, i.e. that which is put into effect by its own bodily organs, is already—in nature and through the force of nature—a true and proper formalization of a generative logic that is open and free. Thought is an acentric activity, or perhaps better, "with a diffuse centre", multidimensional, compresent, intuitive of the whole in the parts and of the parts in the whole. Discourse, however, whether spoken or written, rushes like a flood of signs, granular, sequential, definite, quantifiable, so to speak, as with computers. In some true sense, discourse is a natural anticipation of that digitization with which sounds, colours and figures are now transmitted through spaces. From one person to another the convoy of signs starts from one understanding and arrives at another. Both are, so to speak, already provided with an identical (or almost identical) conversion table or connection or compatibility between sign and concept, between a physical vibration of a sign and an interior vibration of a mind. The indefinite possibilities of different architectures of the distribution of an always limited number of single signs have allowed Rabelais and Leopardi, Dante and

D'Annunzio, Raphael and Picasso, Vivaldi and Wagner, to send us such different messages.

But the situation is different with man and his external computer. Creative, logical, free, intuitive understanding does not exist in a computer, any more than life or psyche, which exist in animals and plants, but without the understanding that is found in human beings. You know how in Catholic teaching it is a dogma that each soul is created immediately by God, i.e. how it needed nothing less than God in His entirety to be able first to work out the designs and then to put them into a "*forma substantialis hominis*", or rather that consistency of being and that force of thinking that characterize each person who alone knows how to speak.

The meaning, then, of the word "spirit" is not one that has been either intellectually constructed or recognized somewhere else and then attributed to human behaviour. It is derived from the unique verification of the existence in us of "a wind", or inner stimuli, to safety, completeness, independence, to dominating and possessing not only the present, but also the future and the unknown and everything. Spirit is opposed to "material", not as an antagonist, but as a manipulator of natural resources and a ruler of nature. Beethoven and Verdi are not the opposites of the orchestra. In the computer, then, native human understanding does not exist. Is it possible to make the computer simulate it? In other words, would it be possible to program a computer which, after scanning a text, would then perform some of those operations as a human being who had understood the meaning of the text? Jacques de Chabannes, Lord of La Palice, would reply "of course! certainly, to the extent that the programmer were able to load into the computer the more or less identical sign-sense correspondence tables that were operative in the mind of the person who wrote that text". Nonetheless, the matter is not quite so self-evident.

Above all, in fact, in the computer's algorithms and memories the sign-sense table can only exist as a sign-sign table, i.e. a table of correspondences between two signs. Yet, so as not to stretch the matter out indefinitely, of the two signs the one cannot be the sign of the sign and the other the sign of the sense. Rather, one is the sign of the verbal sign of the text, and the other is another sign that commands those operations pre-established by the programmer in one of the various informatic approaches that a reasonable person would carry out after reading the text. Furthermore, such a table would have to contain not only the correspondences between signs or groups of elementary signs, but also those of the most probable different types of combinations of words within the most probable types of formats of sentences. The overall meaning of a sentence is not in fact given by the simple sum of the words that are in it, but by the words in that particular arrangement. It would just take a "no" out of place to change the meaning radically.

Lastly, perhaps it would be better to distinguish texts of more complicated hermeneutics from the simpler ones. As is obvious, the hermeneutics of texts of former ages are more difficult, as are, amongst synchronic texts, those of texts that move from the concrete to the abstract, from discourse proper to metaphorical discourse and to rhetorical figures, or indeed from direct discourse to grammatical meta-discourse. However, I am amazed how, if I am not wrong, the elaboration of informatic linguistics turns predominantly on literary, philosophical, or legal texts, rather than on "operative" texts, such as commercial or technical ones.

At this point I will close. Nonetheless I believe that I have cleared up a point, that the internal process of understanding by reading is the corresponding mirror image of expressing oneself in writing. To the extent that this is true, I am of the opinion that a simulation of understanding by reading, for which so many courageously strive, will not come about before the parameters of the variations of our expressive structures are "deprogrammed" from great quantities of natural texts—initially operative ones and then cognitive ones, first homogeneous ones and then different ones. I take the opportunity to repeat my old comparison. An inductive micro-analysis that arises from censuses of the discourse to the internal elementary algorithms with which that discourse is expressed includes and leads us to revise completely—not in order to destroy, but to specify, set out in detail, widen and deepen—our definitions of categories, elements and textual structures, as much as the strategies of informatics demand. That which we have traditionally thought of as morphology, syntax, lexicon and rhetoric, is still an insufficient interface for the use of the computer.

Chapter 17 Considering Myself as if I were a Computer

Roberto Busa S.J.

First published as: Busa, R. 1995. Considerando me come se fossi un computer. *AEI* 82(11) (Nov.): 979–983. [This paper was presented at the *Conferenza di Umanesimo* "Leonardo da Vinci", Milan, 23 May 1995].[1]

Editors' note

Busa delivered this paper in connection with the Leonardo project, which sought to make the case for the importance of the inclusion of humanistic learning in the education of engineers and architects. In it we see Busa reflect on some of the topics that regularly appear in his later articles, for example, the author/work relationship and the nature of the interrelationship between knowledge, words and things. Here he examines these concepts through the prism of the "two cultures" divide. He states his belief that the humanistic and scientific-technological disciplines are fundamentally different yet argues that they overlap in their attention to "man inasmuch as he is an operator on nature". He proceeds to compare himself with a computer in order to isolate that characteristic which is unique to humans. He finds that "A human being is generated by nature, while every machine by definition is produced by man". Busa reflects on the implications of this for the two cultures from various perspectives. For example, he proposes that just as words and knowledge are the manifestations of thought, the things that man creates can be seen in the same way. The human sciences study "thought" and "expression" and so he asks in what way it will be necessary for those disciplines that deal with "things", and for the "sciences and technologies", to similarly attend to "the logic and discourse" that we use in them. In subsequent sections of this paper Busa also calls for attention to "logical ontology" and the "unity of knowledge".

In the search for knowledge, the two cultures (humanistic and scientific-technological) represent two areas that are mentally distinct, but have one overlapping zone, namely the knowledge of man inasmuch as he is an operator on nature. The unity of knowledge is equivalent to the unity of the universe as known by us, and of any "other" universe that can be known. The complementarity among the mathematical sciences, and that of nature, and the human sciences, can be noticed in the way that the former study and describe man who knows, speaks and acts, while the latter express what he knows, what he speaks about, and what he produces.

I will give a didactic cut to my reflections on the integrative complementarities of the two cultures, the humanistic and the scientific-technological: how and in what sense does the latter need the former in making and communicating itself? In comparing myself with the computer, I have tried to find at least one characteristic

[1] [Editors' note] We have omitted from this translation a section on the Leonardo project that was written by Marco Somalvico.

J. Nyhan and M. Passarotti (eds.), *One Origin of Digital Humanities*,
https://doi.org/10.1007/978-3-030-18313-4_16

that, taken in its primary and proper sense, would always and only be found in a human being and never in a machine, or vice-versa, always and only in a machine and never in a human being. In fact, in transferred, i.e. metaphorical usage, with all its implications of similarities and analogies, only existence and nothingness would be free from any sort of unification. I did find one characteristic, although doubtless not the only one that there is, a human being is generated by nature, while every machine by definition is produced by man [Figure 17.1].

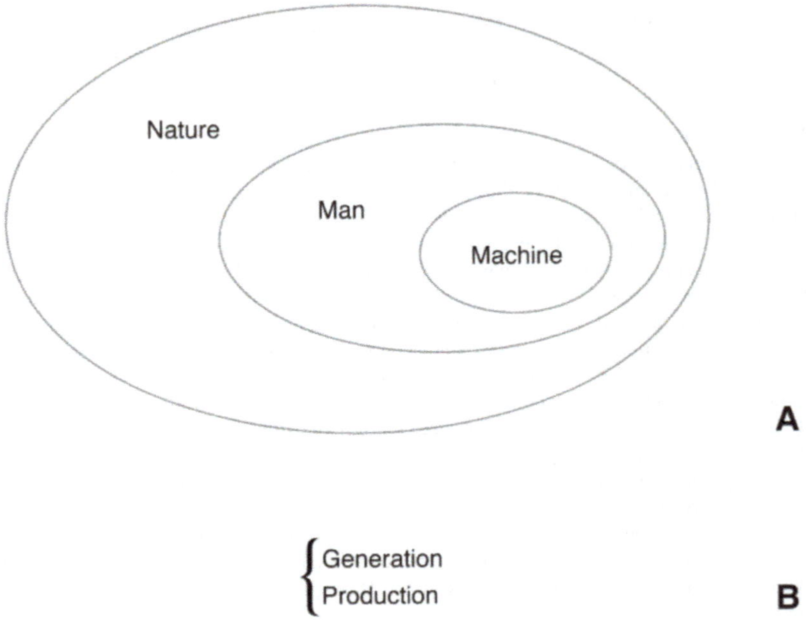

Fig. 17.1. A human being is generated by nature, while every machine by definition is produced by man.

A human being is born in the spaces of nature, while every machine is born only in the spaces of man. The contrast between generating and producing [Figure 17.1] is extraordinary: when parents place their order for a child, they do not make it, but have it made. They do not first draw the blueprint. The child will not be the fruit of their ingenuity. The relation of father to son is not that of author to work. But when a human being produces a machine, to the extent that it is the fruit of his ingenuity, he has the patent as well as the copyright.

In table [17.3], I present my observation in the shape of a proposition, the three parts of which give, from three different angles, an explanation which is actually the same. It is that thought of man, which is expressed by the production of knowledge and words, is also expressed by the production of things. In consequence all technology is also produced, at least empirically or rather through the instinctive physiology of nature, by logic and by expression. Now, we essentially call human sciences exactly those reflective sciences with which we study our thought and our expression. Hence, I put this question: in what ways and in what measures will it be useful, not to say absolutely necessary, to have also in sciences and technologies a carefully considered awareness of the logic and the discourse that we use in them? I see three levels in the logic: that of the including/included (and therefore excluding/excluded), that of the active/passive, and that basic one of the principle of non-contradiction (being/not being).

At the first level I include all mathematics and all geometry, and I stress that the fundamental value of the computer lies in making possible any type of calculation from great masses of data in a very short time. The active/passive relation (causing to be made/being made) is nothing but the operative arch, whose two extremities are always conjoined, heteroclite and irreducible, as are author and work, and, for example, giving and taking, killing and being killed ...

Proposition:	Question:
Sciences and technologies, are to human sciences as works are to their authors	Will it be advantageous in sciences and technologies to advance from the empirical and practiced use of logic and language to an informed training in them *ad hoc*?

Fig. 17.3. A human's thought is expressed with the production of knowledge and words

The author-work arch

Figure [17.4] and [17.5] show the first phases essential to every productive process, but the scheme is easily adaptable to any type of artistic production. Ever since the world began, we have noticed in ourselves three levels, almost as if they were machines of activity and work: knowing, expressing oneself, and every other sort of acting and doing. Within knowledge (Figure [17.4]) there is above all the fact of the first ignition of all curiosity and every "why?", or rather that understanding that there is something which I still do not know and which I need in order to "explain clearly" what I do know.

There follows an acquisition of items of knowledge which in its own way verifies the author/work relation, i.e. when I acquire explanations, pieces of information, formulas ("virtual reality" or rather image-patterns), which I accumulate in the portfolio of my memory, to avail myself of them—when necessary—in manipulating the "real realities" for my own use. For example, the various specializations of medicine and surgery are the fruit of laborious "de-programming" of the physiological and pathological behaviours of different components of our body. What I mean by "de-programming" is the examination of the *output* of a particular behaviour so as to go back and trace the internal logics by which it was produced, i.e. to reconstruct its programme on the basis of the *output*. (I do not say it is an easy task. And from there to identify its author would be a further step, i.e. not an immediate one.)

Level 1: To know philosophy-sciences; to express = to make my thought exist in something else [level 1 overarches every other level]
Level 2a: with the mind; with the senses [level 2a overarches level 3]
Level 2b: *expression of ideas, humanism: in the sense of "the beautiful"; *production of technology
Level 3: *to seek; *to apprehend, to de-program; *to create = to assemble;

Fig. 17.4. Scheme of items of knowledge and expressions

Figure [17.5] shows only three grades or types of production:

- that of items of knowledge (of nature), which are mental images that exist within us as representations of something else, or rather, concepts and thoughts or ideas and formulas, which are realities internal to that activity of ours that we call knowing: "I know what other things are";
- that of elocutive or linguistic expressions (they are not the only ones, but they are the main or fundamental ones) that are entities or physical events produced by us as "signs" in order to communicate the items of our knowledge, transitory but attendant, such as the physical frequencies of our phonations, established on physical supports for use from a distance, such as our writings;
- that of objects manufactured by us, i.e. the "machines" from mincing machines to networks of satellites.

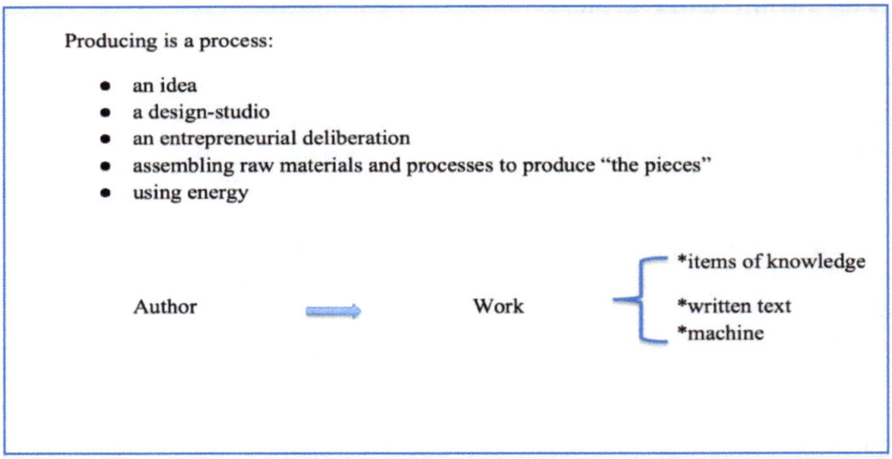

Fig. 17.5. Essential phases of every productive process

But to come back to [Figure 17.4], this shows how the last two types of production can, if need be, unite under the common title of expression, i.e. of communication to others or through others, with the help of supports "external to us" which we have not brought into being and which will continue to exist without us. On the other hand, the fruit of the production, or rather acquisition of items of knowledge of everything else, is a power of acting that is internal to us.

One *exit*, which I will not take, would be to ask oneself where the first idea of something new came from? Was it from full or empty? Was it from chance or by chance? Or is it possible to try to make it spurt out from chance, as a composer does at the keyboard or a pupil when tested in class?

Another *exit* would be to ask ourselves if the mental creativity that we exercise every we time we organize something new—this happens to everyone, if only in our everyday conversations—can simply be reduced to new assemblages of already known elements, or if prime and simple elements can also be "invented" by us.

A third *exit* would be to spend a few minutes analysing the functions and values of sensitive knowledge. In fact, one sensation is presented as the meeting of two operating systems, both of them physical, but one of them is also "psychic". We all know *tout court* how to use our senses, but who knows how to sound their micro-complexity? St Thomas asked himself here and there why we feel matter as a continuous extension... However, at the moment in [Figure 17.4], my only concern is to localize the "human sciences", as sciences that reflect on how we think and how we express ourselves:

> *si come secondo raggio suole*
> *uscir del primo e risalire in suso*
> *pur come peregrine che tornar vuole...*
> *(Dante, Par.1, 49s)*
> (as a second ray comes from the first and goes back up
> just like a peregrine falcon that wants to return...)

If I shine a beam of light into the darkness of a cellar, I acquire some knowledge, but if I shine it at 90 degrees onto a mirror, the reflected ray goes back along itself to "observe" itself, as if an emigrant were to turn back to the house from which he had set out. All humanism starts from "know thyself". Certainly, the two cultures, the humanistic and the scientific-technological, are very different mental areas, but not as different as history and chemistry; they have in fact an overlapping zone, which is the knowledge of man inasmuch as he is an operator on nature.

Sciences and technologies certainly have no need of literature, aesthetical criticism, whether it be literary, pictorial, or musical, neither have they any need of textual criticism or palaeography. Nonetheless, their value is in proportion to the extent that they are the fruit of valid logic, and the extent of their valid communication is in proportion to how appropriately they are expressed in speech and writing. The cultural and social movement called humanism first arose (and continues—I hope I am not wrong here) aiming to systematize in every human expression, first the harmonies of the beautiful, and secondly the holistic proportional setting of each one, tempered to the whole of the vitalities of the complete man, just as today, for example, one speaks of cities made to the measure of man.

Sciences, however, aim at the truth of their own knowledge of nature and technologies to make good use of it.

These are the different aims that distinguish the different spaces of culture. Indeed, they must always remain different, but not separated. Science and technology necessarily incorporate some portion of humanism. For this reason, I mentioned an informed training *ad hoc* in [Figure 17.3] for the use of logic and language: can knowledge be acquired and can machines be constructed without the use of logic, and without talking and writing about it? In St Thomas I have a warning that is most appropriate here: *studium philosophiae non est ad hoc quod sciatur quid homines senserint sed qualiter se habeat veritas rerum* (it is one thing to know the truth of things, quite another to know how man has expressed it). That said, a minimal conclusion of this first chapter could be that of instituting courses for the linguistic analysis of texts and lexicon proper to every single science, pure and applied.

If there were not an (external) observer

I remember, when it was explained to me when I was young, the mental tangles into which that Heisenberg's principle of indetermination threw me. I wrongly connected with the hypothesis of the external observer. Today I have the same thoughts again when I read contemporary articles of cosmic physics about the "anthropic principle" (1). In [Figure 17.2] I have caricatured the two types of "other" that recur in human discourse. There is an "other" which for me is alien, and which by chance or through arrogance penetrates the confines of my vital space. On the other hand, there is an "other" which is "mine", because I have devised it and brought it into being, as in the case of the inventor of the CD-ROM, whose name I cannot recall, only that he was a retired employee of the firm Philips.

Fig. 17.2. Caricature of two types of "other" that recur in human discourse.

The person who knows an object best is its inventor. In sensorial perception, also in that mediated by instruments, every external observer disturbs, i.e. alters in some degree, the situation of which he should like to acquire the absolute image, i.e. that which would have been had he not observed it. That makes sense, because psychic cognitive observation, i.e. sensorial, is always activated by intermediary stimuli, which differ according to the different senses, as for example the light which in a microscope I shine onto the slide. My perception is external to the pulsations of the amoeba that I study, it does not "centre", it does not cause or introduce anything intrinsic there, except perhaps a little bit of "noise". All perception is a casual yet directed system of two operators, the perceiving and the perceived. The key will be sought in the higher system which absorbs the encounters, even the chance ones, of its two lower systems.

Yet the hypothesis stated in the title of this paragraph would not have any proper meaning, if by that we wished to exclude from the reality which is studied the internal observer also, or, above all, the logic incorporated in and, so to speak, superimposed on those realities that operate together. An "idea" that must have come from somewhere, from someone who has programmed its execution. The logic of such a structure is not perceived but produced by such an internal observer, i.e. by the person who knew how to shape and assemble its pieces.

When St Thomas wrote the dry words *omne compositum causam habet* (every compound has a cause) he simply meant to say that every organization must first have been thought out.

This picture meets no objections in the human world: no-one refuses to pay the architect after having paid the building firm for all the work and the suppliers for all the materials.

If some joker, after looking at human neurons through a microscope, were to say that he could not see thought there, he would be sufficiently answered by the retort that thought was not under the slide. Rather, it was above it, behind the eye of the person peering through the microscope, and that the inventor was paid for the microscope, as was the laboratory, research centre and research strategy. Nonetheless, it appears that some may fancy they are capable of thinking that "mere" cosmic evolution over milliards of millennia, and in the space of milliards of millions of light-years, by virtue of an infinite and eternal chance probabilistic combination, has resulted in the coagulation of the great masses of infra-atomic cosmic elements in their differentiations that we call the "laws of nature".

This, however, is an *exit* I will not take, at least not here and now. But if these words had any meaning, it would be that the global logic of the whole cosmos was "no logic". If there is no pre-logic, it is unnecessary to speak of an Author of the cosmos and of us ... (But is it really true that the realization of the probabilities follows a non-logic?) Nonetheless, I think it is important to point out that Christian thought has admitted the facts of the evolution of the species (not of mankind), I think at least from the time of F. Redi, who died in 1698, obviously within other logical frameworks. From about the VIIth century the hymn of Thursday Vespers in the Roman Latin Breviary (today in weeks one and three) has been sung in cathedrals and cloisters. St Thomas alludes to it, and it says:

Magnae Deus potentiae
qui fertile natos aqua
partim relinquis gurgit
partim levas in aera

ut, stirpe ab una prodita,
diversa repleant loca...

(God of great power, you who, of those born of fertile water, leave one part of them in the sea (i.e. leave one part of them fish) and another part you lift into the air (i.e. you transform them into birds), so that, although derived from the same stock, they may fill different places).

If the first part of this paper concluded with the need for a little specific linguistic analysis of scientific lexicon and discourse, the second part concludes—I hope logically—with the need, even in science subjects at university, for some attention to ontology, the hygienic study of our logic. This, in fact, subtends all our scientific acquisition, and consequently the expressions of it. Frankly—and I apologize if anyone dislikes the fact—such ontology should not, because it could not, be distilled from the ontologies of professional philosophies, and much less from the kaleidoscopical history as eminent men have expressed it.

The ontology that is useful to the researcher and the constructor could only have as a basis the lowest common denominator of the logic of imagining, of expressing and bringing into being, which operates vitally in every human being that knows how to speak, whether learned or illiterate, whether more or less intelligent. In this way I have brought you back to the analysis of the expression that concluded my previous section: logical ontology. This lies, in fact, a little above, or a little below, logical analysis and grammatical analysis.

Thus, in conclusion, it remains true that exact sciences and technologies have never done without expressive logic, in the sense that, at least empirically, they have always at least exercised it, through a law of nature. At this point the question might be asked if to teach people to be aware of it would promise qualitative improvements in the acquisition and teaching of science, both pure and applied?

The unity of knowledge

I went to look up this phrase in the *Enciclopedia Einaudi*, whose indices were elaborated on the computer in masterly fashion by two friends, Renato Betti and Piero Slocovich. In vol. XIV, p. 574 *et seq*. there is the summary scheme, which still recalls that idealism which, in Italy too, permeated culture when I was young. I give in my own way the same synthesis in [Figure 17.6].

Internal unity of what is known	* the cosmos
Internal unity of the person who knows[2]	* by type
	* by person
	* by communication amongst those who know
Unity "of system"	* between the person who knows and what is known[3]

Fig. 17.6. The unity of knowledge

And, first of all, the unity of what is known is equivalent to the unity of the universe known by us, and of any "other" universe that can be known. There is a pluralistic unity of the many beings that exist. There is taxonomic unity of their types scanned in bunches within the realms of nature. There is the dynamic unity of what brings into being and what is brought into being, producer and produced, or rather—if such words mean anything—cause and effect. Perhaps it is not possible to say that such a unity has a centre, unless perhaps in the sense of a centre spread everywhere, which would be a fine oxymoron. It is not exactly true that the Middle Ages placed God at the centre of everything (just as it is not true the Giuseppe Verdi is at the centre of Aida). In fact, in the Middle Ages God was considered the first beginning and ultimate end.

The unity of the wise man, or rather of the power of knowing and being aware, is obvious in the first stage: all human beings are alike in the way that—although they do so in such a different fashion—they have knowledge and they think. It is also obvious in the third stage, as they are social communications. But we are not accustomed to the second stage today. There is a small work of St Thomas, "*De unitate intellectus contra Averroistas*" (On the unity of the intellect against the Averroists), in which he champions the idea that there are as many intelligences as there are persons. Who of us ever thought the contrary? Yet it appears that at the time some foremost Arab thinkers, continuators of Greek philosophy, were so struck by the noise of "understanding" that they spoke of it as one single thought of one single being, as if, as we might say today, of an electric generator on a network without transformers. Yet recent absolute idealism spoke of the thought-thinker. When I was a student, there was an amusing parody, in bombastic but de-

[2] [Editors' note]: In Italian: *Unità interna del sapiente*.

[3] [Editors' note]: In Italian: *Tra sapiente e saputo*.

cent terms, that I will quote merely because it sticks in my memory: "*la palingenetica obliterazione dell'io cosciente che s'infutura nell'archetipo dell'antropomorfismo universale ...*" (the palingenetic obliteration of the conscious ego that will futurize itself in the archetype of universal anthropomorphism ...). I do not know who wrote it.

The third unity, according to which knower and known make up an operative system, each one its own, differentiated in its infinite realizations, is that which was the hinge on which I have seen turn the complementarity and integration of, on the one hand, mathematical and natural sciences and, on the other, human sciences. The latter, in fact, study and describe the human being, who knows, speaks and acts, while the former express what he knows, what he speaks about, and what he produces.

Chapter 18 Doing Philosophy on the Computer and Doing Philosophy with the Computer

Roberto Busa S.J.

First published as: Busa, R. 1996. Fare filosofia sul computer e fare filosofia con il computer. *Filosofia e Informatica – Atti del primo incontro italiano sulle applicazioni informatiche e multimediali nelle disc. filosofiche, Nov. 1995, Roma,* ed. Luciano Floridi, 79–84. Paravia.

Editors' note

In this text Busa explores some of the contributions that computing can make to philosophy. He writes that the use of the computer in informatics research gives rise to a particular kind of philosophy that is distinct from that pursued by professional philosophers. Yet, the observations that he goes on to make about language being the starting point of philosophy imply that his remarks are not limited to one kind of philosophy over another. He again argues that the most important contribution of the computer to scholarship is its ability to handle an unprecedented mass of data. Because of this the computer can allow intricate linguistic analysis of any kind of human expression, whether philosophical or not. Hence, he writes that the computer will contribute to philosophy to the extent that words are used to express philosophical thoughts. Busa emphasizes his vast experience of computer-mediated linguistic analysis. He estimates that he had analysed 22 million words in 19 languages and 8 different alphabets by the time of the writing of this article.

Busa then returns, once more, to the author/work relationship. After setting out the steps that are involved in doing, or inventing something, and emphasizing that it is the human being who "had the idea of something new or of some new service" who is entitled to "title and copyright", he envisages that some of the steps involved in moving an idea from conception to instantiation may eventually be delegated to a computer. The portrait of the computer that he paints in these later articles is of a more advanced machine than that of the "high speed cretin" of earlier articles. Yet, he is still at pains to emphasize that which the computer cannot do, and he states his belief that a computer will never be able to invent as humankind can.

1 Philosophy by and on the computer

I should have preferred to rewrite this paper after hearing all the excellent papers given at this conference. I say rewrite, for the topic has been the subject of discussion for over forty years, as will be seen from the bibliography. It appeared, in fact, around the year 1950, when cybernetics "exploded", as education by computer. The word "cybernetics" has today been superseded by "informatics", which, although different as a concept, is nonetheless compatible and no less appropriate.

J. Nyhan and M. Passarotti (eds.), *One Origin of Digital Humanities*,
https://doi.org/10.1007/978-3-030-18313-4_17

The former involves the managing of processes, the latter connotes the organization of information.

Here, therefore, I will confine myself to summarizing the conclusions, which have already been published, of my experiences of linguistic analysis performed on 11 million words of philosophical texts in mediaeval Latin and on the same number of words of texts of different literary genres in 18 other languages and 8 different alphabets. Moreover, it will be clear from what I say, why I acknowledge philosophical nature also as one part of the linguistic analysis of the latter.

1.1 Philosophies of the "mind-body" relationship

After the strident journalistic rhetoric of the 1950s was replaced by calm scientific methodologies, the need for a definition of how to place the use of electronic instruments within a universal scheme of the whole, has led mankind—within the spaces of the man/machine relationship—to enquire whether there are boundaries (and, if so, what sort of boundaries, whether combinatorial with what is above, or "participatory" with what is below) between thought and brain, mind and body, spirit and material. And this seems to be a philosophy that arises from the factual data of the technologies of informatics. In other words, it seems that it does not flourish through internal evolution of the living "professional" philosophies, except inasmuch as a common culture is imbued with them. Obviously in such a philosophy one can see the ridge, as it were, between two slopes, namely, between materialism and spiritualism, or, to put it in a nutshell, between atheism and prayer.

1.2 Man and machine are the author-work unit of doing and causing to be done

In my writings I start from the position that man and machine are neither rivals nor enemies, except, for example, when in real life a machine that is alien to him produces a noise or smoke that is harmful to him, or, to take another example, makes copies of books protected by copyright. Properly speaking, however, man and machine confirm the author-work relationship. Each piece of technology is one of the human expressions and its glory. Every computer "is a system", i.e. it is a unit that operates with the person who produced it, the person who programmed it, and the person who uses it.

There are, so to speak, machines for work, such as, for example, engines, with which man applies energies of nature to act on surrounding bodies for his service. The computer, too, is such a machine, but its work consists in elaborating physical entities—the bits—which have been adopted by man to act as pieces of information, i.e. as signs of his own knowledge. In the communicative interface that is called "word and discourse", both written and oral, the sign is, in fact, external to the man, i.e. it is something other than he, inserted by him to exist or subsist in the surrounding bodies, while the meaning or concept exists and remains only in the mind of the speaker or writer and of the hearer or reader. Every system is an organization, an assembly, of already existing forces put together to collaborate hierarchically for the production of a specific result or manufactured article.

Indeed, it seems that amongst the animals only man is an operator capable of "inventing", i.e. continually creating for himself new instruments for new purposes for his own advantage. Machines, technologies and computers exist only in human spaces, as an offline extension of the forces of man's hands and of his mind. Only man knows how not only to do but also how to cause to be done. "Doing" implies above all:

a) having had the idea of something new or of some new service;
b) describing its executive designs in detail;
c) entrepreneurial deliberation;
d) providing the raw materials;
e) lastly, organizing the workforce to "construct" it.

Title and copyright are granted to the person who had the idea and developed the design.

For this reason, children are not the "work" of their parents: once the order for them has been sent out, the parents "allow" them to be made by nature. They have not made the designs by which three thousand billion cells are produced in nine months and so differentiated amongst themselves as to form a living, bawling baby that weighs around 2.5 kg at birth. "Causing to be done" implies knowing how to delegate to others at least one of the functions "b, d, c". I certainly think it is possible to delegate, in some degree, function "c" to a computer, or rather that of deliberating on the execution of something. Yet I cannot imagine that function "a" can ever be delegated: to delegate "being the first to" seems to me a contradiction in terms—one can say the words but not think them to be true. When one speaks of "a delegate", is that not also to speak of "a second"?

1.3 Information and form

So, in the man-computer system, the sign is in the computer, the meaning is in the mind of man. And it is precisely because the operative sign-sense combination is called "information" that other ways to philosophy on the computer have been opened up, to the extent that the word "information" is extended even to programs intrinsic to nature. In fact, the phrase "information theory" is used in mathematics, and the word "information" is used with reference to the genetic, biochemical, molecular and atomic structures of DNA.

Again, within the author-work relationship, noticing that the "contents" of our knowledge are the "forms" of reality, gives us an inkling that information and the thought that has either conceived it or recognized it are essentially connected, the first for efficiency, the second for destination. When the "forms", or the formulas of the structures, or the logics of the programs incorporated in things, are established in our knowledge, we call them ideas, concepts, representations, formulas, models, images, possibilities... "appearances, phenomena, virtual realities...". These are values of power enabling us to move ourselves and produce something in our life. Form and individual, nature and singularity, possibility or existence, essence and being, thought and reality, are not these the heights of philosophical speculation? Aristotle used to say that every form, seen in this way, has something of the divine and of the dynamic (1).

2 Is the computer useful for doing philosophy?

This question would be banal if it were no more than to ask how writing or printing machines could "compose" texts of philosophy. Yet it would become a page of science-fiction, perhaps even an entertaining one, if it were elevated to an examination of how something manufactured by a human being could "think" philosophy. But if one asks seriously in what ways and measures a computer could be useful for doing philosophy, in my opinion the reply would have to take four considerations into account.

2.1 The three original phases of philosophizing

In the phrase "doing philosophy" I see the following as three levels, or stages, or sections:

a) conceiving of it or thinking of it;
b) expressing it, or formulating its words, internally to oneself;
c) expressing it to others, i.e. communicating it, publishing it, teaching it.

"b" and "c" are also a first "formalization" of our thought, that of those expressive signs that are the words of our language, formed spontaneously through social instinct.

2.2 Either to discover for oneself or to understand by listening and reading

There are two ways to acquire knowledge: one is comprehension, or intuition, or discovery, from direct personal experience. The other is learning from someone else's discourse (2). The first is the way of inventions and of "originality", the second is the fruit of communication. And to communicate implies repetition of knowledge. But let it be granted that communication, both oral and written, is not passed down, but maieutic: he who communicates does not inject his understanding into another, but occasions in the other, or just merely favours, the birth of understanding within him.

2.3 In discourse: expressive vehicle and expressed message

In all discourse two tiers are to be distinguished: the expressive vehicle and the expressed message. With the latter "that which" is meant is specified; the former is "that with which" it is expressed. Generally speaking, the expression-vehicle is constituted by:

a) the morphology of the words and the syntax of the sentences;
b) the so-called grammatical words, i.e. prepositions, conjunctions, pronouns, and auxiliary and modal verbs.

All other verbs, common nouns and proper nouns, specify the message.

However, through the different types and topics of communication, this latter part of the vocabulary—in synchrony—varies appreciably, while the former does not change in equally conspicuous measures. Within communication, both at the points of departure and arrival, the expressive vehicle expresses that logic of our mind which in it is the reflection of being, of acting, and of doing. These are the universal values of reality, common, that is, to every different mode, sector and stratum of the cosmos and of life. In my writings I call this "generative ontology", the spark that ignites all our dynamism of knowing, communicating and living.

2.4 What is "doing philosophy"?

What does "doing philosophy" mean? What is philosophy? I do not think it is an easy question, nor do I think that an answer is often to be found. I will give the answer that seems to me to be the correct one, knowing that it can be formulated in many different ways, which are also compatible. Philosophy is a process of thought that prepares and brings to maturity a certain type, a certain fascia of our cognitive operations which have the "*reductio ad unitatem*" as a specific content, i.e. which succeed in grasping the systematic unitarity of the whole. The fact that there are very many wholes is a derivation and an expression of a fullness of form and power. The two currents of logic that bring such a process about are the including/included relationship for the form and the active/passive relationship for existence and activity.

And that is to be understood in the sense that in the "very many wholes" are included not only the reality of bodies external to our knowledge (I carefully avoid saying "external to our body") but also that other reality which is our speech, as well as that other reality which is our knowledge and our thinking. These activities of ours are realities which, open to the horizon, emerge from our bodies just as a radar periscope emerges from the hull of a submarine. The *reduction ad unitatem*, or rather a reply to our last "whys", is therefore also the ancient "*via interioritatis*", i.e. it is a reflection with which we understand ourselves in the activity of thinking. The famous phrase *gnothi seauton* inscribed in the temple at Delphi could be understood as a synonym for philosophy: know yourself, seek out your roots.

Philosophy, inasmuch as it is a social inheritance of the human community, can only start from the analysis of how common it is to all in speaking. Speaking, in fact, is the interface that socially connects cosmos and thought. If one were to study the cosmos only, one would only know one's own body, from the atoms to the animal psyche. One would know the kingdom but not the king. In terms of quantity our body is insignificant: the whole of the present human race, if piled up together, would not fill a cube of one kilometre .... But our thought is the dynamically terrible scholar, thirsting to dominate everything, and forever.

I will not go too deeply into one last observation: philosophy starts from the examination of speech, but it's very last and vital goal is not that of formulating either many grand words or one single infinite word, but of finding it, recognizing it and loving it. And loving is definitely not the same as knowing and saying.

2.5 The computer also assists philosophy in the analysis-synthesis of the speech of everyone and of everything.

From the foregoing remarks I conclude that the computer will participate in the philosophical mental processes in the proportion in which words play a part in them as physical signs. And I have not forgotten two facts: first, that the computer too—as by definition does every instrument—enables man to do much more than he would have been able to do just with his own body. Secondly, that the computer's own contribution is to sift the high mountain of the enormous and to dredge the deep river of the occasional and casual, in short to dominate great masses of elementary elements.

That electronic, even multimedia publishing, databases and webs of databases can promulgate philosophical texts is obvious, and that in return they cause increases in the didactics of philosophy (that there are two steps in making it memorized and making it understood is also obvious). Moreover, if it is true that philosophy and the history of philosophy are distinct like the two cotyledons of the same seed (3), it is not difficult to imagine what sort of services the computer could render to the history of philosophy.

But let us try to place ourselves exactly in the path from "a" to "b", as mentioned in 2.1. above. It is a two-way journey. One journey is, from having gradually intuited concepts of synthesis, to ascend to expressing them with words, first to oneself and then to others. The second journey is to rise from the reading or hearing of someone else's discourses to bring them gradually to maturity within oneself. What help can informatics give to these original and essential moments of doing philosophy?

In the observation that the goad of the "whys", which activates all research, is nothing but the exercise of that same logic of being and causing to be, of which morphology, syntax and lexicon are the expressive reflection, I find the confirmation of how much my publications have repeatedly affirmed that. It makes sense to speak of philosophy helped by the computer—just as it makes sense to speak about computerized hermeneutics—

- because it makes sense to speak of linguistic analysis and synthesis on the computer and
- because the analysis-synthesis of our speech is the very runway from which philosophy takes off.

In fact, to study one's own words and those of someone else, in one's own and in other languages, leads one to identify what sort of thought, at least at a basic level, formed, regulated and chose them.

In philosophical texts a great part of the specific message will obviously be philosophical. But in every text or discourse, including non-philosophical ones, the analysis of the vehicular material of its language will always lead to a verification of its philosophical nature. Thanks to the computer, linguistic analysis-synthesis can today achieve levels of scientific rigour that were previously impossible. There is the analysis of the single words of which the publications of the *Lessico Intellettuale Europeo* are the model.

There is also the analytic and synthetic description of the entire vocabulary of a written work or corpus, a map of the lexicological system operating there. I carried this out for the *Index Thomisticus* and for many texts of other languages. With it I have always gathered and synthesized every word without exception, at least on the two levels of forms of words that are graphically different, and of the lemmas, i.e. of the lexical units to which the individual forms must be retraced. I sometimes also grouped into thematic families all the forms of all the lemmas constructed around one same theme.

Thus, my reply to the question that I asked myself is as follows: the computer is a help also for philosophy inasmuch as it allows censuses, analyses and syntheses of all the factors, elements and linguistic categories of any human expression, both that which is specifically philosophical and that which deals with something else. And "from the point to which thought shall have brought the computer, from there thought shall continue on its own" ... I took a more recent new proof of this from the research that I published in my last book *Inquisitiones Lexicologicae*. This was an investigation into the semantic heterogeneity of words. It concluded with giving quantities and percentages to the distribution, in those eleven million words, of the basic categories of reality. That is, it ended by being a synthesis of the last internal structures of "things and words and concepts", in correspondence—apart from the general organization—with both the "*praedicamenta*" of Aristotle and the categories of Kant.

2.6 A fortiori the computer assists "sacred" theology

What I have affirmed for philosophy, I consider to hold true—and in some sense *a fortiori*—for theology. I mean the "sacred" or revealed theology, which is the hermeneutics of how much God personally lets us know about Him and about the whole in Holy Scripture, as seen in His dialogue—that is in the vicissitudes of time—with the Church of His Christ. "Natural" theology, however, is the last chapter of the philosophy which concludes all the preceding ones, studying there the signs of the presence of God and, from these, His prerogatives.

Now, in linguistic terms the difference between sacred theology and philosophy is the following: philosophy interprets that exclusive discourse of God which is the life of nature in time. In fact, only God can create without pre-existing material something that is not Him but from Him, as His expression. Theology, however, interprets the human and historical language with which God has revealed to mankind both His ineffable mysteries, and what future awaits them outside of time.

Bibliography

1. In 1973 Andre Robinet of the CNRS in Paris started the periodical CIRPHO (*Centre International Recherches Philosophiques par Ordinateur*) in joint publication with McGill University in Montreal; it ceased in 1976.
   In Rome the *Lessico Intellettuale Europeo* project of the CNR, founded and directed by Tullio Gregory, does "computer assisted" philosophy much more than it theorizes about it. Amongst its various activities it organizes international conferences every three years on single words that are purely philosophical, and it publishes the proceedings. The

words that have been discussed are: *ordo, res, spiritus, imaginatio-phantasia, idea, ratio* and *sensus*. These words can also be found in my publications on Aquinas.

2. The following have commented on my linguistic analyses:

A. Di Maio, L'Informatica Linguistica di p. Roberto Busa come methodo investigativo e come approccio al Medioevo, *Medioevo*, 1 (1989), pp. 325–362.
R. Diodato, Tra linguistica e ontologia: tipi di semanticità emergenti dal lessico tomista, *Rivista di Filosofia Neo-scolastica*, 83(4), [1991] pp. 512–524.
P. Ponzio, La tradizione filologia nella linguistica computazionale: la vicenda di Roberto Busa SJ, in *Congresso Annuale A/CA 22–24 Settembre 1993, Gallipoli, (LE)*, pp. 627–636.
C. Toulon-Galli, Interview de Roberto Busa SJ – Histoire de L'Index Thomisticus – Un demi siècle d'informatique linguistique, in 3eme *Colloque Histoire de L'Informatique*, Sophia Antipolis 1993, Pre-conference Proceedings.
P. Guietti, Hermeneutics of Aquinas' Texts: Notes on the Index Thomisticus, in *The Thomist*, 57–4 (Oct. 1993) pp. 667–686.

3. What I have written above derives from those of my publications that are listed below: the initial number indicates their place in the complete list.

012 La Cibernetica, in *Ingegneria Meccanica*, 3–2 (Feb. 1954) pp. 5–10.
018 Cibernetica, in *Enciclopedia Filosofica*, Florence 1957, 1, 1022–1025.
028 I principali problemi dell'automazione del linguaggio scritto in *Atti della VI Sessione delle Giornate della Scienza, Convegno Int. Sui Problemi dell'Automatismo, Mila, 8–13 April 1956, Rome 1958,* vol. 1. pp. 832–841.
032 L'automation appliqué a l'analyse linguistique des ouvrages de Saint Thomas d'Aquin: programme, etat actuel, in *L'homme et son destin d'apres les penseurs du Moyen Age, Actes du 1er Congres Int, de Philos. Medievale. 1958*, Louvain 1960, pp. 619–625.
055 Saggio esplorativo di automatic abstracting, in *Convegno Nazionale T.D. 66: La Riduzione Concettuale dei Documenti, Turin, 14 Dec. 1966*. Proceedings edited by CSAO, Turin (1967), pp. 39–47.
056 Funzionamento e uso di un calcolatore elettronico, in *L'uomo e la macchina. Atti del XXI Congr. Naz. Di Filosofia. Pisa 22–25 April 1967*, Turin 1967, vol. iii pp. 240–245.
057 Article in *L'uomo e la macchina. Proceedings of the XXIst Congr. Naz. Di Filosofia. Pisa 22–25 April 1967*, Turin 1967, vol. III pp. 259–263.
077 Cibernetica: macchina e uomo, in *La Civilta Cattolica*, 2859–2860 (2 and 16 August 1969) pp.272–275. (Press review).
087 L'Index Thomisticus e L'Informatica Filosofica, in *Revue Internationale de Philosophie*, 103 (1973–1) pp. 31–36.
092 Le implicazioni filosofiche della *Cibernetica, in Humanismo Pluridimensional – Atas da Primeira Semana Intern. De Filosofia, Sao Paulo 16–22 July 1972*, Sao Paulo, Brazil, 1974, 2 vols, pp. 607–624.
096 L'Index Thomisticus. Documentazione computerizzata per l'analisi linguistic dell'opera di San Tommaso d'Aquino, in *L'Osservatore Romano*, First part n. 205 (34,685) 7 Sept. 1974 p. 5. Second part n. 207 (34,687) 9–120 Sept. 1974 p. 5.
097 L'attualità dell'impostazione linguistica che San Tommaso dà alla filosofia in Presença Filosofica, Sao Paulo Brazil 1974, 1–2–3, pp. 186–194. Reprint of the second part of the article in *Osservatore Romano* v. 096.
105 Inner and Outer Information – Causality types in man-to-man information, in *Proceedings of the Jurema 1975 – Intern. Symposium on Cybernetics in modern Science and Society*, Zagreb 1975, pp. 17–20.

106 The computer in Theological Research, in *Fachsprache – Umgangssprache, Kronberg*/Ts 1975 pp. 403–424.
189 Busa R., Biffi I., L'Index Thomisticus: per la filosofia e la teologia, in *Teologia (Rivista Fac. Teol. It. Sett.)*, Morcelliana (Sett. 1980), pp. 257–271.
193 Modelli del conoscere e del parlare in Il sapere come rete di modelli – La conoscenza oggi, Modena 1981, pp. 235–244.
210 L'analisi linguistica del linguaggio dei mistici – Prospettiva di elaborazione elettronica degli scritti di S. Veronica Giuliani, in *Testimonianza e Messaggio di Santa Veronica Giuliani – Atti del Congr. Int. Roma Pont. At. Antonianum 27–31 Oct. 1982*, Rome 1983, 2 vols pp. 466+400, in vol. 1 pp. 129–134.
219 L'ontologie générative chez S. Thomas d'Aquin, in *L'homme et son univers au Moyen Age – Actes 7eme Congres Int. Philosophie Medievale, 30 August – 4 Sept. 1982*, Louvain-La-Neuve 1986, pp. XIII = 961 pp. 496–504.
225 On Law and Freedom – Hermeneutical Methods for Interpreting St. Thomas from the Index Thomisticus in Lex et Libertas (...), *Proc. Fourth Symp. on St. Thomas Aq. Philosophy, Rolduc 8–9 Nov. 1986*, Pont. Acc. S. Tommaso 1987 pp. 25–40.
229 *L'originalite Linguistique de St. Thomas d'Aquin,* in *A.L.M.A.* 44–45, (1985) pp. 66–90.
233 Fondamenti di Informatica Linguistica, Milan 1987.
234 Das Problem der Thomistischen Hermeneutik nach der Veroeffentlichung des Index Thomisticus, in *Miscellanea Medievalia*, 19 (1988) pp. 359–364.
237 Il Lessico Intellettuale Europeo, in *La Civilta Cattolica*, 3314 (16–7–88) pp. 151–157.
262 Far e Far Fare: Uomo e Macchina, in *Intelligenza natural e Intelligenza artificiale – XLIII Conv. St. Fil. di Gallarate – Apr. 1988*, Genoa 1991 pp. 57–77.
263 L'analisi del parlare in S. Tommaso: nuove prospettive di ermeneutica tomistica, in *Atti del IX Congr. Tomist. Int., Roma 24–29 Sept. 1990, Rome 1991* pp. 212–223.
265 De expressione apud S. Thomam in Littera Sensu Sententia, – *studi in onore di Cl. J. Van Steenkiste OP*, Milan 1991 pp. 135–154.
277 Nouvelles Perspectives d' Hermeneutique Thomiste, in *Methodologies Informatiques et Nouveaux Horizons dans les Recherches Medievales,* Brepols 1992 pp. 65–76.
278 L'interiorità principio di tutto: traguardo di ogni filosofia, in *Interiorità (...) Contributi al XLV Conv. Centro St. Filosofici di Gallarate, April 1990*, Genoa 1992 pp. 75–83.
294 Algoritmi interiori del capire leggendo in *Macchine per leggere. Tradizioni e nuove tecnologie per capire i testi. Atti (...) Certosa del Galluzzo 199.11.93*, Spoleto 1994 pp. 69–75.
296 Inquisitiones Lexicologicae in Indicem Thomisticum, 2a editio emendata auctaque, CAEL 31–5–1994 pp. 218.
304 L'Enciclopedia Multimediale delle Scienze Filosofiche, *La Civilta Cattolica* n. 3481 (1 July 1995) pp. 52–55.

1. I Phys. 9, 192a 166–17: *(morphes) ontos gar tinos theiou kai agathou kai ephetou*. St Thomas mentions these 5 times in 4 works: in the *Summa c. Gentiles* 005-SCG lb 1 cp50 n. 9 and lb3 cp69 n. 27; in the *In De Caelo et Mundo* 050-CCM lb3 lc2 n.2, *In libros Perihermeneias* 058 lb1 lc3 n.8; obviously in the *in libros Physicorum* 052-CPY lb1 lc15 n.7.
2. "*Causa semper excedit causatum. In quibusdam autem locutio causat intellectum (here it means comprehension, concept) sicut in his quae per disciplinam (teaching) discuntur (...) Quandoque autem intellectus est causa locutionis, sicut in his quae per inventionem sciuntur.*" St Thomas, *In 4 libros Sententiarum* 001–1SN ds37 qu2 ar3 *expositio textus.*

3. St Thomas again in the *In De caelo et Mundo* 050-CCM lb1 lc22 n. 8: "*studium philosophiae non est ad hoc ut sciatur quid senserint homines, sed qualiter se habeat veritas rerum*".

Chapter 19 Roberto Busa S.J. Bibliography: 1949–2009

Editors' note

This bibliography has been derived from a file that was originally drawn up by Busa and that he subsequently circulated among colleagues and friends. We have extended this original file and corrected its incomplete or inaccurate bibliographical references. We have also omitted a number of the publications that are included in it because they are texts about Busa, or published by his centre, rather than texts that he himself authored. As the Editors view interviews as a format that is co-created by interviewer and interviewee, we have included here the references to interviews in which Busa participated.

1949

La *Terminologia Tomistica dell'Interiorità. Saggi di metodo per una interpretazione della metafisica della presenza*. Milano: Fratelli Bocca [pp. 280.].

A proposito di immanenza e trascendenza. In *Congreso Intern. de Filosofia. Barcelona, 4–10 Oct. 1948. Actas.* Vol. II, 671–689. Madrid

1950

Complete Index Verborum of St. Thomas Aquinas. *Speculum – A Journal of Mediaeval Studies* 25(3), 424–425.

1951

S.Thomae Aquinatis Hymnorum Ritualium Varia Specimina Concordantiarum. Primo saggio di indici di parole automaticamente composti e stampati da macchine IBM a schede perforate. Milano: Bocca.

Rapida e meccanica composizione e pubblicazione di indici e concordanze di parole mediante macchine elettrocontabili. *Aevum* XXV(6), 479–493.

1952

Rapidissima composizione di Indici e Concordanze di parole mediante schede perforate. In *La documentazione in Italia. Atti del XVIII Congresso Mondiale di Documentazione [...]*, 95–97. Roma: Cons. Naz. delle Ricerche.

Mechanisierung der philologischen Analyse. *Nachrichten fuer Dokumentation* 3(1), 14–19.

Review of: S. Scimè – Il trionfo dell'ontologismo in Sicilia: Giuseppe Romano. *Archivum Historicum Societas Jesu* 1(41), 210–212.

1953

Le Concordanze Lombardiane. *Pier Lombardo* 1–2, 15–18.

Entwicklungen der Mechanisierung der Sprachlichen Analyse. *Nachrichten für Dokumentation* 4(4), 202–204.

1954

Libri per i piccoli. *Letture* IX(1), Milano, 16–19.

J. Nyhan and M. Passarotti (eds.), *One Origin of Digital Humanities*,
https://doi.org/10.1007/978-3-030-18313-4_18

La Cibernetica. *Ingegneria Meccanica* III(2), 5–10.

La Cibernetica. *IBM World Trade News*. Ott. 8 s.; Nov. p. 5; Dic. p. 11.

1955

L'Index Thomisticus. In *Sapientia Aquinatis. Communicationes IV Congr. Thom. Intern*, 602–605. Roma.

1956

La Filosofia di Pier Lombardo. *Miscellanea Lombardiana*, 33–44. Novara: De Agostini,

1957

Die Elektronentechnik in der Mechanisierung der sprachwissenschaftlichen Analyse. *Nachrichten für Dokumentation* 7(1), p. 7.

In *Enciclopedia Filosofica*. Firenze: Sansoni:

- *Cibernetica*, I, 1022–1025.
- *Concorso Divino*, I, 1162–1170.
- *Differenza e Diversità*, I, 1568–1570.
- R. Busa & C. Negro, *Dio*, I, 1594–1610.
- *Distinzione*, I, 1657–1672.
- *Divisione*, I, 1687–1694.
- R. Busa & C. Negro, *Finalità, principio di*, II, 438–439.
- *Lippert, Peter*, III, 87–88.

1958

The Use of Punched Cards in Linguistic Analysis. In *Punched Cards – Their Application to Science and Industry*, 2nd edition, ed. R.S. Casey, James W. Perry, Madeline M. Berry and Allen Kent, 357–373. New York: Reinhold.

The Index of all non-Biblical Sea Scrolls published up to December 1957. *Revue de Qumran* 1(2), 187–198.

I principali problemi dell'automazione del linguaggio scritto. In *Atti della VI Sessione delle Giornate della Scienza – Convegno Int. sui Problemi dell'Automatismo, Milano, 8 – 13 Aprile 1956*, vol. I 831–841. Roma: Consiglio Nazionale delle Ricerche.

1959

All non-biblical Dead Sea Scrolls published up to December 1957 have been indexed. In *Sacra Pagina, Miscellanea Biblica Congressus Internationalis de Re Biblica (1958)*, 7–12. Paris: Lecoffre.

1960

Zusammengefasste Darstellung der Erfahrungen des Centro per l'Automazione dell'Analisi Letteraria des Aloisianum. *Kolloquium Maschinelle Methoden der literarischen Analyse und der Lexicographie*, 24–26, 6+26 Tuebingen [litograph].

Erlaeuterungen zu den lexicographischen Arbeiten zu Goethe, Farbenlehre, Bd. 3. Ibidem, 36.

L'automation appliquée à l'analyse linguistique des ouvrages de Saint Thomas d'Aquin: programme, état actuel. In *L'homme et son destin d'après les penseurs du Moyen Age. Actes du 1° Congrés Int. de Philos. Médiévale. 1958*, 619–625. Louvain: Nauwelaerts.

Saggio di derivazione della Teologia Naturale dalla Problematica del Consenso Universale sulla Divinità. *Miscellanea Gazzana* II, 329–348. Milano.

1961

Les travaux du Centro per l'Automazione dell'Analisi Letteraria de Gallarate. *Cahiers de Lexicologie* 3, 64–70.

L'analyse des homographes et des formes flexionnelles, ibidem, 134–139.

1962

Atti del Convegno: Linguistica e Industria Oggi, ed R. Busa, 75. Milano [mimeograph].

Notes on some lexical and philosophical implications of a computer symbolic language. In *Symbolic Languages in Data Processing* – Proceedings of the Symposium organized and edited by the ICC. Rome, March 26–31, 759–761. New York: Gordon and Breach.

R. Busa, C. Croatto-Martinolli, L. Croatto, C. Tagliavini & A. Zampolli. Una ricerca statistica sulla composizione fonologica della lingua italiana parlata eseguita con un sistema IBM a schede perforate. In *Proceedings of the XXIth Intern. Speech and Voice Therapy Conference, Padua 1962*, 542–562.

L'Automazione dell'Analisi Lessicale. In *La meccanizzazione delle attività linguistiche. Atti del Convegno tenuto a Torino presso la Camera di Commercio Industria e Agricoltura il 27–9–1962*, 27–30.

Notes bibliographiques et notions sur les aspects humanistes, philosophiques et religieux de l'émigration. In *Migrations* II(3–4), 67–74 [pp. 27, 28, 29, 41].

L'Analisi linguistica nell'evoluzione mondiale dei mezzi d'informazione. In *Almanacco Letterario Bompiani 1962*, ed. Sergio Morando, 103–107. Milano: Bompiani.

Stellungnahme. In *Zur mechanischen Sprachuebersetzung. J. J. BECHER, Allgemeine Verschluesselung der Sprachen* von Prof. Dr. W. G. V. Waffenschmidt, 46–52. Stuttgart: Kohlhammer Verlag [A publication of the Wirtschaftshochschule Mannheim, r. 1: pamphlet, b. 10].

1963

L'automazione dell'indagine lessicale sui testi di Qumran. In *Atti del Convegno La Lombardia e l'Oriente, 11–15 giugno 1962*, 60–65. Milano: Ist. Lomb. Accad. di Scienze e Lettere.

1964

De Linguae Latinae Opportunitate et usu ad Scientias per automationem euulgandas. In *3me Congrès Int. pour le Latin Vivant, Strasbourg, 2–4 Septembre 1963*, 119–121. Avignon: Aubanel.

La Cibernetica sta provocando una vera rivoluzione industriale. *La Stampa* (quot.) 18 Sett. 1964, Torino.

1965

An inventory of fifteen million words. In *Literary Data Processing Conference Proceedings, Sept. 9–10–11, 1964,* eds. Jess B. Bessinger Jr. et al., 64–78. New York: Modern Language Assoc.

Review of: S. S. Acquaviva – Automazione e Nuova Classe. Bologna 1964, *La Civiltà Cattolica* quad. 2769 (6–XI), 272–273.

World Evolution in Information Processing and its Influence on Linguistic Research. In *Third World Congress of Jewish Studies. Jerusalem 1961,* 47–49. Jerusalem [in Hebrew].

1966

Die Elektronentechnik in der Mechanisierung der sprachwissenschaftlichen Analyse. In *Avtomatizatsiva v lingvistike*, L. N. Zasorina, Nauka: Moskow [a translation into Russian of the 1957 article above with the same title].

Nota de nominibus Dei in Lexico S. Thomae. In De Deo [...] Acta VI Congr. Thomistici Internationalis, Romae, *Doctor Communis* XIX–XX (1966–1967) 83–88.

La Cibernetica e le possibilità dell'uomo nuovo. *Informazione Scientifica* XII, n. 510, 22–26; n. 511 p. 23–27.

La Cibernetica e le possibilità dell'uomo nuovo. *Il Fuoco* n. 3, (Maggio-Giugno), 19–33. Roma.

Risultati ed esperienze di fatto sui preparativi previi all'impiego del calcolo automatico in biologia. In *Atti del Convegno Internaz. sulla Sclerosi Multipla, Gallarate 24–25, Aprile 1966,* ed. R. Montanini, 377–382. Gallarate.

1967

Saggio esplorativo di automatic abstracting. In *Convegno Nazionale T.D. 66: La Riduzione Concettuale dei Documenti. Torino, 14 Dic. 1966. Atti*, 39–47. Torino: CSAO.

Funzionamento e uso di un calcolatore elettronico. In *L'uomo e la macchina. Atti del XXI Congr. Naz. di Filosofia. Pisa 22–25 Aprile 1967*, vol. III, 240–245. Torino: Edizioni di Filosofia.

Contribution: ibidem, 259–263 del vol. III.

In *Enciclopedia Filosofica*, 2a edizione, Sansoni:

- *Cibernetica,* I coll. 1387–1391.
- *Concorso Divino*, I, coll. 1559–1568.
- *Differenza e Diversità, II, coll. 448–449.*
- W. Brugger & R. Busa, *Dio*, II, coll. 470–492.
- *Distinzione*, II, coll. 552–564.

- *Divisione*, II, coll. 582–588.
- *Finalità, principio di*, II, coll. 140–1401.
- *Lippert, Peter*, IV, coll. 3–4.

1968

Validità della prova della esistenza di Dio. In *Problemi Teologici d'Oggi*, 21–28. Milano: Uff. Catechistico Dioc.

Cibernetica del Linguaggio. In *Atti e Rassegna Tecnica della Società degli Ingegneri e degli Architetti in Torino* 22(1), 1–3.

De Lexico Electronico Latino. In *Acta Omnium Gentium ac Nationum Conventus Latinis Litteris (...) aprilis a. MDCCCCLXVI Romae habiti*, 383–410. Roma: C. Colombo.

Un lexique latin électronique. In *Les Machines dans la Linguistique. Colloque Intern. sur la mécanisation et l'automation des recherches linguistiques*, 251–269. Prague: Acad. Tchécoslovaque des Sciences.

R. Busa & A. Zampolli, *Centre pour l'Automation de l'Analyse Linguistique (C.A.A.L.), Gallarate*. ibidem, 25–34. Traitement des mots d'une fréquence extreme. ibidem, 137–142.

Erreurs Humaines dans la Préparation de l'Input pour Ordinateurs. ibidem, 279–284.

L'Index Thomisticus della Facoltà di Filosofia dell'Aloisianum di Gallarate. In *L'elaborazione elettronica nell'analisi linguistica,* 1–23 con 25 tav. Pisa: Centro Naz. Univ. Calc. Elettron., 29 + tav. 33 [litograph].

Indicis Thomistici edendi rationes, mensurae atque specimina. Pisa: Centro Naz. Univ. Calc. Elettronico [litograph; 50 + tav. 37].

Actes du Séminaire International sur le Dictionnaire Latin de Machine, Pisa 27–29 mars 1968. Ed. R. Busa, *Calcolo* 5, uscito come supplemento n. 2, 176. INAC: Roma.

1969

Actes du Séminaire International sur le Dictionnaire Latin de Machine. *Revue* (Organ. Intern. Etude Langues Anciennes par Ordinateur) 2 and 3, 176. Liege, Belgique.

Cibernetica: macchina e uomo. In *La Civiltà Cattolica*. a. 120, 2859–2860, 272–275.

1970

L'Instrumentation électronique dans les recherches linguistiques. In *IV Rencontre Int. de Méchanographie et Informatique, Lisbonne Oct. 1967*, 737–747. Caixa Nacional de Pensòes.

Gli sviluppi della Cibernetica. In Bollettino del Rotary Club La Malpensa (Busto Arsizio) 5 (1968–1970), 111–113.

1971

Concordances. In *Encyclopedia of Library and Information Science* vol. 5°, 592–604. Marcel Dekker: New York.

Review of: Jagjit Singh – Teoria dell'Informazione. Linguaggio e Cibernetica. Milano. Mondadori, 1970, 8°, 357. In *La Civiltà Cattolica* 122(2911), 95–96.

1972

The Impact of Cybernetics on the Humanities. In *Proceedings of the Jurema 1972 – International Symposium on Cybernetics in Modern Science and Society*, ed. Prof. Vl. Muljevic, 13–24. Elektrotechnicki Fakultet: Zagreb.

Problemi operativi del trattamento elettronico di quantità enormi di testi naturali. In *Rivista Trimestrale di Scienza dell'Amministrazione*, XIX(1), 50–58 [translated from the French by E. Riccato].

Review of: Irma Zorzi, Le sere così lente, Padova, Rebellato, 1969, 66. In *Letture* 8–9, 64–71, Milano.

Il lessico di san Tommaso – intervista con padre Roberto Busa. In *Rivista IBM* VIII(3), 16–21. Milano [interview].

1973

Index Thomisticus. Il "computer" al servizio degli studi tomistici. In *L'Osservatore Romano* pag. 5, 20 Genn., n. 15 (34.196)

L'Index Thomisticus e l'Informatica Filosofica. *Revue Internationale de Philosophie* 103, 31–36.

A Venezia gli uomini peggio del mare, recensione di: A. Zorzi, Venezia scomparsa 1972, 2 vols. In *Letture*, 6–7/73, 497–499. Milano.

Review of: Lessico di Frequenza della lingua italiana contemporanea, di U. Bortolini, C. Tagliavini, A. Zampolli, 1972. In *Letture*, 12/73, 843–845. Milano,

Review of: Cristina Brambilla, La Negritudine, 1972. In *Letture*, 12/73, 847–848. Milano.

1974

[See below for a list of the parts of the *Index Thomisticus* that were published from 1974–80].

L'Index Thomisticus – Contenuto, Finalità, Prospettive. In *La Civiltà Cattolica,* a. 125, 2967 (2 Febbraio), 250–257.

Le implicazioni filosofiche della Cibernetica, 607–624; Index Thomisticus. Un esempio di elaborazione elettronica in campo umanistico. In *Humanismo Pluridimensional – Atas da Primeira Semana Intern. de Filosofia, Sao Paulo 16–22 Julho 1972*, 2 vols, 1036: 625–636. Sao Paulo Brazil: Loyola.

Associationi inquirentium per automaton de scriptis hominum sermonibus ipsa suae rei natura docet fines alios statui non posse quam qui internationalitatis. *Assoc. for Literary and Linguistic Computing Bulletin* 2(2), 2–3. [guest editorial].

L'Index Thomisticus. Documentazione computerizzata per l'analisi linguistica dell'opera di San Tommaso d'Aquino. *L'Osservatore Romano* Prima parte, 7 Sett. 1974, n. 205 (34.685) p. 5. Seconda, 9–10 Sett. 1974, n. 207 (34.687) p. 5.

L'attualità dell'impostazione linguistica che San Tommaso dà alla filosofia. In *Presença Filosofica* 1–2–3, 186–194. Sao Paulo: Brazil [a reprint of the second part of the article in *Osservatore Romano* above].

Review of: Silvio Tramontin: Carità o Giustizia? Torino, Marietti 1973. In *Letture* 12–74, 867–868. Milano.

L'esistenza di Dio per l'uomo d'oggi. In *Hum-anismo Pluridimensional – Atas da Primeira Semana Intern. de Filosofia, Sao Paulo 16–22 Julho 1972*, 684–689. Sao Paulo Brazil: Loyola.

1975

ALLC 1974 Annual General Meeting. *Assoc. for Literary and Linguistic Computing Bulletin* 3(1), 55–56. [Text of an address that Busa gave to the meeting].

L'esplorazione dell'infinito: dalla cibernetica al pensiero. *Il Fuoco* 23(1), 20–24 & 33. Roma.

Inner and Outer Information – Causality types in man-to-man information. In *Proceedings of the Jurema 1975 – Intern. Symposium on Cybernetics in modern Science and Society*, ed. W. Muljevic, 17–20. Zagreb.

The computer in Theological Research. *Fachsprache – Umgangssprache*, eds. J.S. Petoefi, A. Podlech, E. von Savigny, 403–424. Scriptor Verlag: Kronberg/Ts.

Der Index Thomisticus. *IBM Nachrichten* 228, 317–324. Stuttgart.

R. Busa & A. Zampolli, *Concordantiae Senecanae*. Georg Olms: Hildesheim [2 vols, I – A – Naturam IX + 1–822; V II – Naturam – Zonas + Indices, ibidem VII + 823–1473 + 176].

1976

Review of: Silvio Tramontin, Sinistra Cattolica di Ieri e di Oggi, Torino, Marietti, 1974, pp. 182. In *Letture* 31–325, 252–253. Milano.

L'Index Thomisticus e le Metodologie delle Ricerche Tomistiche. In *Atti del Congr. Intern. (Roma-Napoli, 17–24 Aprile 1974) Tommaso d'Aquino nel suo Settimo Centenario*, vol. II 432–437. Napoli: Domenicane Italiane.

Why can a computer do so little? *ALLC Bulletin* 4(1), 1–3 [Guest Editorial].

Review of: Mimi Zorzi, La nuova età, Venezia-Padova, Marsilio, 1975, 160. In *Letture* 31–327, 371–374. Milano.

Interventi 82, 222s, 232, 276, 294; risposte ai questionari App. I 309–350. In *1. Colloquio Internazionale del Lessico Intellettuale Europeo*, (Roma 7–9 Gennaio 1974), eds. M. Fattori & M.L. Bianchi. Roma: ediz. Ateneo.

Review of: F. Rivetti Barbò, Semantica bidimensionale. Fondazione filosofica, con un progetto di teoria del significato, ed. Elia, Roma 1974 476. In *Rivista di Filosofia Neoscolastica,* LXVIII, fasc. II, 333–337.

Computer Processing of over Ten Million Words: Retrospective Criticism. In *The Computer in Literary and Linguistic Studies – Proceedings of the Third International Symposium*, eds. Alan Jones & R. F. Churchhouse, 114–117. Cardiff: The Univ. of Wales Press.

The quantities of the Latin vocabulary documented in the Index Thomisticus. *Revue* (R.E.L.O.) XII(1–4), p. 1–45.

1977

Cibernetica e Antropologia. *Il Fuoco* 25(2), 5–11, Roma.

The morphological categories of Forcellini's Latin Lemmas. In *Lexicologie – een bundel opstellen voor F. DE TOLLENAERE [...],* 79–106 Groningen: Wolters-Noordhoff b. v.

Review of: Luciano De Crescenzo 'Così parlò Bellavista' 1977. In *Letture* 3(78), 192–193. Milano.

Introduzione alla bibliografia, agli indici e lessici tomistici. In *Seminarium* S. Congr. pro Instit. Cath., Città del Vaticano, n. 3 (nova series a. XVII), 922–958.

1978

Review of: Mario Saggin, Le quattro stagioni di Luca, Marton Ed., 1977, 130. In *Letture* 5(78), 403–404. Milano.

Il linguaggio esercizio d'anima. In *L'Osservatore Romano* CXVIII(128), p. 3.

Epistemologia dell'atto di fede. In *Il senso della filosofia cristiana, oggi – Atti XXXII Conv. Centro Studi Filosofici*, 99–102. Morcelliana: Brescia.

Logica umana dell'informatica. In *Informatica e Documentazione*, Infor-av 4(2), 148–149. Roma.

1979

Review of: Amicizie, di I. Zorzi, Venezia, Marsilio, 1978. In *Letture*, 2/79 128–129. Milano.

Man-machine relationship in computerized linguistics. In *XXVII Convegno Internazionale delle Comunicazioni, Genova 9–12 Ottobre 1979*, Ist. Int. delle Comunicazioni, Villa Piaggio, Genova, 279–286.

Clavis Indicis Thomistici. Stuttgart: Frommann-Holzboog [pp. 48].

ORDO dans les oeuvres de St. Thomas D'Aquin. In *ORDO – II Coll. Intern. Lessico Intellettuale Europeo, Roma 7–9 Gennaio 1977*, 59–184. Roma: ediz. Ateneo-Bizzarri.

1974–1980, **INDEX THOMISTICUS: SANCTI THOMAE AQUINATIS OPERUM OMNIUM INDICES ET CONCORDANTIAE,** in quibus verborum omnium et singulorum formae et lemmata cum suis frequentiis et contextibus variis modis referuntur, quaeque auspice Paulo VI Summo Pontifice, consociata plurium opera atque electronico IBM automato usus, digessit Robertus Busa SI in Gallaratensi Facultate Philosophica Aloisiani Collegii professor. Stuttgart: Frommann Holzboog. 56 voll. ISBN 3 7728 0532 9

SECTIO PRIMA INDI CES

PROSPECTUS DISTRIBUTIONIS

1976 v [1] I – Lemmatum A – Finis / Formarum A – C	pp. XVI + 1064
1976 v [2] II – Formarum D – N	pp. XVI + 1102
1976 v [3] III – Formarum O – Z	pp. XVI + 1038

SINGILLATA DISTRIBUTION

1976 v [4] IV – Lemmatum A – R	pp. XVI + 1099
1976 v [5] V – Lemmatum S – Finis / Formarum A – C	pp. XVI + 1253
1976 v [6] VI – Formarum D – I	pp. XVI + 1259
1976 v [7] VII – Formarum L – P	pp. XVI + 1062
1976 v [8] VIII – Formarum Q – Finis	pp. XVI + 1145

SYSTEMATA LEXICI

1980 v [9] IX – Systema Lexicologicum, Tabulae 1–2	pp. XVI + 1257
1980 v [10] X – Systema Lexicologicum, Tabulae 2–5; Systema Homographiae, Tabulae 6–12; Systema Typologicum, Tabulae 13–26; Systema Quantitatum, Tab.27–38	pp. XIV + 1213

SECTIO SECUNDA CONCORDANTIAE SANCTI THOMAE

CONCORDANTIA PRIMA

1974 v [11] I – A -Agonotheta	pp. XXII + 1140
1974 v [12] II – Agresta – Audientia	pp. XXII + 1126
1974 v [13] III – Audio – Cavea	pp. XXII + 1134
1974 v [14] IV – Caveo – Confisco	pp. XXII + 1143
1974 v [15] V – Confiteor – Corsica	pp. XXII + 1150
1974 v [16] VI – Cortex – Deveho	pp. XXII + 1262
1974 v [17] VII – Devenio Dyrrachium	pp. XXII + 1145
1974 v [18] VIII – E – Facio	pp. XXII + 1120
1974 v [19] IX – Facio(Fio) – Gausape	pp. XXII + 1142
1974 v [20] X – Gaza – Imitor	pp. XXII + 1149
1975 v [21] XI – Immaculatus – Intellectus	pp. XXII + 1237
1975 v [22] XII – Intelligentia – Loquor	pp. XXII + 1170
1975 v [23] XIII – Lora – Moralizo	pp. XXII + 1143
1975 v [24] XIV – Morasthi – Nigromanticus	pp. XXII + 1146
1975 v [25] XV – Nihil – Paricidalis	pp. XXII + 1148
1975 v [26] XVI – Paries – Permanentia	pp. XXII + 1148
1975 v [27] XVII – Permaneo – Praesidium	pp. XXII + 1146
1975 v [28] XVIII – Praesignantia – Quousque	pp. XXII + 1139

1975 v [29] XIX – Rabanus – Sappho	pp. XXII + 1164
1975 v [30] XX – Sara – Specus	pp. XXII + 1167
1975 v [31] XXI – Spelta Tempus	pp. XXII + 1170
1975 v [32] XXII – Temulentia – Vienna	pp. XXII + 1162
1975 v [33] XXIII – Vigeo – Finis	pp. XXII + 1150

CONCORDANTIA ALTERA

1980 v [34] I -A – Deploro	pp. XXIV + 1286
1980 v [35] II – Depono – Hujuscemodi	pp. XVI + 1282
1980 v [36] III – Hujusmodi – Notitia	pp. XVI + 1287
1980 v [37] IV – Noto – Quilibet	pp. XVI + 1293
1980 v [38] V – Quin – Tanto	pp. XVI + 1288
1980 v [39] VI – Tantum – Num. Arab. = Qui	pp. XVI + 1300
1980 v [40] VII – Et = In = Quam quod	pp. XVI + 1270
1980 v [41] VIII – Ad = Non = Sum	pp. XVI + 1297

SECTIOTERTIA CONCORDANTIAE ALIORUM AUCTORUM CONCORDANTIA PRIMA

1980 v [42] I – A – Confiteor	pp. XVI + 1151
1980 v [43] II – Confixio – Ezechiel	pp. XVI + 1150
1980 v [44] III – Faba – Lapis	pp. XVI + 1153
1980 v [45] IV – Lappa – Patientia	pp. XVI + 1146
1980 v [46] V – Patior – Senarius	pp. XVI + 1150
1980 v [47] VI – Senator – Finis	pp. XVI + 1148

CONCORDANTIA ALTERA

1980 v [48] I – A – Mytilena	pp. XVI + 1149
1980 v [49] II -Naaman – Finis	pp. XVI + 1183

SUPPLEMENTUM OPERA OMNIA

1980 v [50] I – In Quattuor Libros Sententiarum	pp. VIII + 724
1980 v [51] II – Summa contra Gentiles, Autographi Deleta, Summa Theologiae	pp. VIII + 952
1980 v [52] III – Quaestiones Disputatae, Quaestiones Quodlibetales, Opuscula	pp. VIII + 668
1980 v [53] IV – Commentaria in Aristotelem et alios	pp. VIII + 600
1980 v [54] V – Commentaria in Scripturas	pp. VIII + 528
1980 v [55] VI – Reportationes, Opuscula dubiae authenticitatis	pp. VIII + 612
1980 v [56] VII – Aliorum Medii Aevi Auctorum Scripta 61	pp. VIII + 840

1980

The Annals of Humanities Computing: the Index Thomisticus. *Computers and the Humanities* 14(2), 83–90.

R Busa & M. Nicolodi. The Annals of Humanities Computing: the Index Thomisticus. *Medicina nei Secoli* XVII(3) (Sept.–Dec), Univ. Roma, 443–459 [text translated into Italian from the English (see above) by M. Nicolodi].

P. Busa ha completato la pubblicazione più grande della seconda metà del secolo. *La Prealpina* (Quotid. Varese) 93–271 (Sab. 22 Nov.) p. 23 [interview].

1981

R Busa and I Biffi. L'Index Thomisticus: per la filosofia e la teologia. *Teologia* (Rivista Fac. Teol. It. Sett.), 257–271. Morcelliana.

L'Index Thomisticus. *Studi Medievali* XXI(1) 3a serie, 411–421. Spoleto.

Per S. Tommaso 'ratio seminalis' significa 'codice genetico' – problemi e metodi di lessicologia e lessicografia tomistiche. In *Atti dell'VIII Congr. Tomistico Intern.* vol. 1, 437–451. Città del Vaticano.

Modelli del conoscere e del parlare. In *Il sapere come rete di modelli – La conoscenza oggi* (Convegno Intern. Modena 20–23 Gennaio 1981), 235–244. Modena: ediz. Panini.

Contribution: in *Atti della Giornata di Studio sul Vocabolario Giuridico Italiano, Firenze, 26 Genn. 1979,* (Ist. Documentazione Giuridica del CNR), 88–90.

Una filosofia nuova ma sempre antica, recensione di: G. Sommavilla, Il pensiero non è un labirinto – Dialettica e Mistero, Jaca Book. In *Letture* 12, 905–908. Milano; and in *La Civiltà Cattolica* 3151–3156, 571–574.

R Busa R & M Nicolodi, Il progetto e la realizzazione dell'Index Thomisticus. In *La Scuola e l'Uomo (UCIIM, Roma)* XXXVIII(10) (Ott), 233–238. [The same Italian text that Nicoldi translated from the 1980 'Annals' article above].

Conversazioni con padre Roberto Busa: S. Tommaso: Opere, Indici, Concordanze, *L'Osservatore Romano* CXXI(123) (29–30 Maggio) p. 3.; L'Index: Inventario d'un Sistema Linguistico, *L'Osservatore Romano* CXXI(124) (31 Maggio) p. 3.

Review of: Mimi Zorzi: Il medico di famiglia, Milano 1981. *Letture* 8–9, (Ag.–Sett.) 575–576.

Il mistero del linguaggio. In I. *Biffi – C. Marabelli, Invito al Medioevo, Conversazioni*, 57–64. Jaca Book: Milano.

1982

Voces Realis-Realiter. In S. Thoma Aq. cum appendice de voce Res-Rei, 105–136; Activités depuis 1977, 507–508, in *Lessico Intellettuale Europeo – Res, 3. Coll. Intern., Roma, 7–9 Genn. 1980, Atti.* Roma: ediz. Ateneo.

De Vocum Frequentiis in Usum Lexicologiae Priusquam Stylisticae Inducendis. In *L'Analisi Delle Frequenze – Problemi di Lessicologia* (Less. Intellett. Europeo – Roma 7–9 Genn. 1980), 111–122. Roma: ediz. Ateneo.

Global Linguistic Statistical Methods to Locate Style Identities – Proceedings of an International Seminar, Gallarate, June 5–7 1981, ed. R. Busa (L ess. Intell. Europeo XXIX). Roma: Ateneo [pp. 111].

1983

Trent'anni d'informatica su testi: a che punto siamo? Quali spazi aperti alla ricerca? In *Conv. su L'Univ. e l'evoluzione delle Tecnologie Informatiche – Milano 14–16 Marzo 1983,* Atti, 2 vols, 7.1–7.4 Milano: CILEA.

Trente ans d'analyse informatique de textes: Où en est-on? Et après?. In *Actes du Congrès Int. Informatique et Sciences Humaines – Liège 18–21 Nov. 198*1, 135–148. Liège: ed. L.A.S.L.A.

L'analisi linguistica del linguaggio dei mistici – Prospettiva di elaborazione elettronica degli scritti di S. Veronica Giuliani. In *Testimonianza e Messaggio di Santa Veronica Giuliani* = Atti del Congr. Int. Roma Pont. At. Antonianum 27–31 Ott. 1982, vol. I 129–134. Roma: Laurentianum.

1984

De voce spiritus in operibus S. Thomae Aquinatis. In *Spiritus – IV Coll. Intern. del Lessico Intellettuale Europeo*, Roma 7–9 Gennaio 1983, 191–222. Roma: ediz. Ateneo.

Educare al conoscere. In *Progetto Scuola* anno II no. 8, 2–6.

L'impiego del computer nei repertori di documenti. In *Archiva Ecclesiae* XXVI–XXVII (1983–1984), pp. 145–148. Città del Vaticano.

1985

De terminationum latinarum statisticis mensuris ex Indice Thomistico. In *Hommage à Pierre Guiraud*, Nice Les Belles Lettres, 147–161.

Lezioni e Documentazioni di Lessicografia e Lessicologia Tomistiche – Gallarate, Milano, Roma (1974–1980). Aloisianum: Gallarate [typescript; compl. 335].

Informatica Linguistica. In *Cultura e Scuola* n. 95, 12–31. Roma: Ist. della Enc. Ital.

Lezioni di Informatica Linguistica. Milano: Università Cattolica [pp. 137 + 50].

Il logos principio d'ogni dialogo tra persone secondo S. Tommaso. In *Congr. del Ventennio dal Conc. Vaticano II- Roma 18–21 Febbr. 1985*, vol. I 577–584. Pont. Università Urbaniana.

De linguae latinae flexivis terminationibus. *Revue Informatique et Statistique dans les Sciences humaines*, XXI(1–4), 53–66.

L'originalité Linguistique de St. Thomas d'Aquin. In *A.L.M.A.* XLIV–XLV, 66–90. Brill: Leiden.

1986

L'ontologie générative chez S. Thomas d' Aquin. In *L'homme et son univers au Moyen Age – Actes 7° Congrés Int. Philosophie Médiévale, 30 Aout-4 Sept. 1982*, 496–504. Louvain-La-Neuve.

Informatica e Nuova Filologia. *Lessicografia, Filologia e Critica – Atti Conv. Int. di Studi, Catania-Siracusa 26–28 Apr. 1985*, 17–25. Firenze: Olschki.

Informatica e Linguistica. *Broteria* 123(4), 310–322.

L'informatica: nuovi orizzonti per la filologia. In *Il foglio e l'albero*, 50–56. Milano.

1987

Un safari nella propria bocca. In *Atti: Giornata di Studio su Informatica e Didattica*

Congresso ann. AICA Palermo 24–24 Sett. 1986, 7.

On Law and Freedom – Hermeneutical Methods for Interpreting St. Thomas from the Index Thomisticus. In *Lex et Libertas [...] Proc. Fourth Symp. on St. Thomas Aq. Philosophy, Rolduc 8–9 Nov. 1986*, 25–40. Pont. Acc. S. Tommaso, ed. Vaticana.

L'interpunzione nelle edizioni computerizzate per L'Index Thomisticus. In *Grafia e Interpunzionenel Medioevo – Semin. Intern. Roma 27–29 Sett. 1984*, 207–214. Roma: Ateneo.

L'intelligenza artificiale: intervista a padre Busa. *il Sabato* X(34), 22 Ag., 19–20 [interview].

Noivado da Filologia e do Computador. *Broteria* 125(4), 256–267.

Fisica e Metafisica dell'algoritmo espressivo. *Il Fuoco* XXXV(3), 6–10. Roma.

Speleologia Linguistica: dati e fatti di segmentazione morfotematica. In *Atti del 3 Conv. Naz. CAI – Torino, Università. 16–18 Giu. 1987*, 179–186. Torino: ediz. La Scientifica.

Se senti bussar apri. E' la logica di Dio. *Il Sabato* 5–11 Dic., p. 33.

Fondamenti di Informatica Linguistica. Milano: Vita e Pensiero. [pp. 412].

Review of: F. Bonicalzi 'Il costruttore di automi. Descartes e le ragioni dell'anima' Jaca Book 1987, *Synesis* IV(3–4), 149–153.

1988

Das Problem der Thomistichen Hermeneutik nach der Veroeffentlichung des Index Thomisticus.

In *Miscellanea Medievalia* vol. 19, 359–364. Koeln: Thomas Inst.

Review of: Francesca Rivetti Barbò 'Dialogo e Malinteso, [...]' Milano 1985. In *Rivista di Filosofia neo-scolastica* LXXIX–3 (Lug.–Sett.), 485–487.

De phantasia et imaginatione iuxta S. Thomam. In *Phantasia-Imaginatio: V Coll. Internaz.*

Roma 9–11 Genn. 1986 – Atti..., 135–152. Roma: Ateneo.

Il Lessico Intellettuale Europeo. In *La Civiltà Cattolica* 3314 (16–7), 151–157.

Um Autor que se apresenta. *Broteria* 127(2) Ag.-Sett., 217–223.

Inteligència natural e inteligència artificial. *Broteria* 127(4) Outubro, 260–268.

Totius Latinitatis Lemmata – quae ex Aeg. Forcellinii Patavina Editione 1940 a fronte, a tergo atque morphologice, opera IBM automati ordinaverat Robertus Busa. Milano: Ist. Lomb. di Sc. e Lett [pp. XVI + 532].

Procédures et résultats de la segmentation thématique des lemmes latins de l'Index Thomisticus.

Revue Informatique et Statistique dans les Sciences humaines XXIV(1–4), 117–132.

1989

Storia Informatica di Parole. *Gregorianum* 70(1), 127–140. Roma.

Contribution: in *Per la lingua di Montale – ATTI (...) Firenze 26 Nov. 1987*, 153–154. Olschki.

Come rintracciare la linguistica di San Tommaso, 37–48; Intelligenza naturale e intelligenza artificiale. In *Homo Loquens*, 272–277. Bologna: ed. Studio Domenicano.

Review of: J.A. Weisheipl 'Tommaso d'Aquino' Jaca Book 1987. *Il Sabato* XII(3) (3 Genn.), p. 80.

Review of: E. Zoffoli. Principii di Filosofia – con letture di storia del pensiero occidentale. Fonti Vive – Cipi Roma 1988. In *La Civiltà Cattolica* 3, 314–315.

La famigliarità della scienza con il paradosso. In *Litterae Communionis* 9, 52–54.

Gli Instrumenta Lexicologica Latina. *La Civiltà Cattolica* 3341, 396–403.

Mes vingt années suivantes. *Revue Informatique et Statistique dans les Sciences humaines* XXV(1–4), 27–34.

1990

Conteùdo da obra 'Totius Latinitatis Lemmata' – (Apliçào da Informàtica linguistica). *Broteria* 130(1), 92–97.

Tutto S. Tommaso d'Aquino su CD-ROM con hypertext. *Informatica* 3/90 XX(3), Lug.-Sett., 273–276. AICA: Milano.

Review of: "Il pellegrino cherubico" di Angelus Silesius. La 'luce essenziale' dell'anima. *Vita Pastorale* LXXVIII(10) (Ott.), 26–27.

Nuova Filologia Computerizzata. In E. Borello, *Linguistica e Comunicazione,* 181–185. Torino.

Il Thesaurus Augustinianus. *La Civiltà Cattolica* n. 3370 (17 Nov.), 351–356.

Informatics and New Philology. *Computers and the Humanities* 24(5–6), 339–343.

1991

Idea negli scritti di Tommaso d'Aquino. In *Idea – VI Coll. Intern. Less. Intell. Europeo, Roma 5–7 Genn. 1989, Atti*, 63–87. Roma: Ateneo.

Fare e Far Fare: Uomo e Macchina. In *Intelligenza naturale e Intelligenza artificiale -*

XLIII Conv. St. Fil. di Gallarate – Apr. 1988, 57–77. Genova: Marietti.

L'analisi del parlare in S. Tommaso: nuove prospettive di ermeneutica tomistica. In *Atti del IX Congr. Tomist. Int., Roma 24–29 Sett. 1990*, vol. I 212–223. Roma: Vaticana.

La Transcendencia de Dios y el Ateismo a la luz del Index Thomisticus. In *3. Congreso Mundial de Filosofia Cristiana – Quito Ecuador 9–14 Julio 1989*, Quito Univ. Cat. Ecuador, 1990–1, vol. 3–4: 147–163.

De expressione apud S. Thomam. In *Littera Sensus Sententia – studi in onore di Cl. J. Van Steenkiste OP*, 135–154. Massimo: Milano.

Préface – Nouvelle Philologie. In *Computational Lexicology and Lexicography [...] dedicated to Bernard Quemada,* Linguistica Computazionale VI–VII, vol. II: IX–XIII. Giardini: Pisa.

Cinquant'anni a 'bitizzar' parole. In *Conv. Int. sulla Storia e Preistoria del Calcolo Automatico e dell'Informatica. Siena 10–12 Sett. 1991*, Atti Precongressuali, 72–82.

Half a Century of Literary Computing: Towards a 'new' Philology. In *Int. Conference on Current Issues in Computational Linguistics, Penang, Malaysia, 11–14 June 1991, Atti Precongressuali*, 84–95.

Scripta Ignatii de Loyola quaecumque extant ab eius nativitate 500mo decurrente anno electronice edita. Gallarate [private edition; 140].

Thomae Aquinatis Opera Omnia – cum hypertextibus – in CD-ROM – auctore Roberto Busa, pp. 64 + 1. Milano: Editel [Compact Disk Read Only Memory].

Thomae Aquinatis Opera Omnia cum hypertextibus in CD-ROM. *Revue Informatique et Statistique dans les Sciences humaines* XXVII(1–4), 95–97.

1992

Half a century of Literary Computing: Towards a 'new' Philology. *Literary and Linguistic Computing* vol. 7(1), 69–73.

Technical Report on the Texts with Hypertexts plus the Lexicological System of the Index Thomisticus on Tapes. May 31. Gallarate: Cael-Aloisianum [pp. 50].

Review of: AA.VV. (a cura di Giovanni Giovannini) Dalla selce al silicio. Storia dei mass media. Torino, Gutenberg 2000, 1999. *Letture* 47(489) (Ag.–Sett.), 680–682.

Mezzo secolo di informatica linguistica. In *Responsabilità del Sapere* vol. 181–182, 3–13. Roma.

Nouvelles Perspectives d'Herméneutique Thomiste. In *Méthod. Informatiques... dans les Recherches Médiévales* (Actes Coll. Int. FIEPM, Saint-Paul-de Vence 3–5 Sept. 1990), 65–76. Brépols.

L'interiorità principio di tutto: traguardo di ogni filosofia. In *Interiorità [...] Contributi al XLV Centro St. Filosofici di Gallarate, Aprile 1990*, 75–83. Genova: Marietti.

The five texts processed beside the Index Thomisticus – Technical Report. Gallarate: CAEL [pp. 23].

Review of: Paolo Rocchi 'L'informazione, i sistemi e il controllo, Mondadori Milano 1992. *Rivista di Informatica* (AICA) XXII(4), p. 393.

1993

La coscienza. In *La Civiltà Cattolica* n. 3430 15(5), 339–350.

Nuove nozze tra Filologia e Mercurio – la formica e la sintesi. In *Il Secondo Rinascimento* 7 (Mag.-Giu.), 161–167. Milano.

Contribution: in *Il Vetrinista* Anno VI Numero 4.

L'ossimoro e la differenza. *Il Secondo Rinascimento* 9, 57–65. Milano.

Thesaurus vocum et ubicationum in Thomae Aq. textibus in CD-ROM. Milano: Editel [pp. 608].

1994

Vocis 'ratio', quae in Thomae Aq. propriis operibus 36.335 vicibus occurrit, rationes atque numeros paucis hic conabar absolvere verbis. In *Ratio – VII Coll. Int. Roma 9–11 Genn. 1992*, 173–195. Olschki.

Umanesimo e informatica ovvero 'bitizzare' parole. *Ingegneria Informatica – guida alla scelta degli studi ...*, 71–73. Politecnico di Milano.

L'umanità a una svolta. Il XIX Congresso Mondiale di Filosofia a Mosca. *La Civiltà Cattolica* 3453 (7 Maggio), 264–270.

Algoritmi interiori del capire leggendo. In *Macchine per leggere. Atti [...] Certosa del Galluzzo 19.11.93*, ed Claudio Leonardi, Marcello Morelli and Francesco Santi, 69–75. Spoleto: Centro It. Studi Alto Medioevo.

Ermeneutica e Traduzione – ... un lessico tomistico 'biculturale'. *Medioevo* XVIII, 3–20.

Inquisitiones Lexicologicae in Indicem Thomisticum. CAEL: Mediolani [Latin and English; 150].

Inquisitiones Lexicologicae in Indicem Thomisticum. A Roberto Busa S.I. latino sermone confectae atque a Philip Barras in anglicum sermonem translatae. 2a ed. emendata auctaque. Gallarate: CAEL [pp. 218].

Padre Busa: intervista a un pioniere – Così San Tommaso finì nel computer. *Rassegna dell'Istruzione* 48(4) (Lug–Ag.), 53–56 [interview].

1995

Review of: Margherita Porete: Lo Specchio delle anime semplici. trad. di Giovanna Fozzer, prefaz. storica di R. Guarnieri, commento di M. Vannini. San Paolo, Cinisello Balsamo 1994. *Vita e Pensiero* 2, 150–152.

Machiavelli e l'Indice dei libri proibiti. In *Il Secondo Rinascimento* 16, 83–86. Milano.

Progetto di una banca dati dei testi di preghiera – Procedure, impegni, problemi e vantaggi di una 'biblioteca elettronica' dei testi mistici. In *Semiotica del Testo Mistico* – Congr. Int., L'Aquila 24–30 Giugno 1991, 414–423. L'Aquila: Gallo Cedrone

Insellare la lessicologia sull'ippogrifo. *AEI* (Ass. Elettrot. Elettron. Ital.) 82–4 (Aprile), 322–328.

Il primato del sensibile. In *30 GIORNI* XIII(5) (Maggio), 28–32. Roma [interview].

La banca del Paradiso. *Il Secondo Rinascimento* 21, 29–31.

Saggezza nel computer. *KOS* Milano 119–120 (Ag. Sett.), 50–53.

Linguaggi naturali. Review of: K. Adel et al. Tiefenstrukturen der Sprache, Wien, Universitaetverlag 1994. *Riv. di Informatica* XXV(3), 219–221

Review of: Pietro di Vona: I concetti trascendenti in Sebastian Izquierdo e nella Scolastica del Seicento – Napoli (Loffredo) 1994, 8. *Archivum Historicum S.J.* 64, 201–203.

Tipos de Semantica en el léxico bì-cultural de Santo Tomas. In *XX Semana Tomistica – Soc. Tomista Argentina, Buenos Aires, 1995*. (Actas). Vers. di M-C. Donadio de Gandolfi.

Considerando me come se fossi un computer. *AEI* 82(11) (Nov.), 979–983.

Realizzazioni e Promesse delle Metodologie Informatiche per una Nuova Stagione di Studi Tomistici. *San Tommaso Filosofo, Pont. Accad. di S. Tommaso*, 371–381. Città del Vaticano.

Review of: Paolo Rocchi – Potremmo "bitizzare" i processi mentali. *Mass Media* 14(5) (Ott.Dic.), 52–53

1996

Preface: *Rodolfo Bozzi S.J.: Introduzione al futuro – Ipotesi sulle radici delle scienze dell'informazione*, 5–13. Bari.

Afterword: *Alessio Petralli: Neologismi e nuovi media*, 119–125. Bologna: CLUEB.

Preface: *A.V. Lingua Letteratura Computer, a cura di Mario Ricciardi*, 9–16. Torino: Bollati Boringhieri.

Mezzo secolo a bitizzar parole. *IF – Riv. Fondaz. IBM Italia* IV(1), 94–99.

The five texts processed beside the Index Thomisticus – Technical Report. 2nd edition. Gallarate, March 7. CAEL: Aloisianum [pp. 25].

Orazione funebre per Mario Piazza sab 23–3. *Rassegna ASLAI* XI(2) (Lug.), p. VII. Brescia.

Thomae Aquinatis Opera Omnia, cum hypertextibus in CD-ROM. 2a ed. Gallarate: Thomistica [compact-disk].

Thomae Aq. Op. Om. c. h. in CD-ROM – Liber Manualis. Milano: Editel [pp. 162].

Il Libro dei Metodi 1: Primi approcci. CAEL: Gallarate [pp. 155].

Goldoni in byte. Review of: L'ipertesto d'autore. La famiglia dell'antiquario su CD-ROM. Marsilio, Venezia 1996. *Il Sole* 24 Ore, Dom. 8 Sett. N246. p.37.

Sensus e Sensatio nell'Index Thomisticus – Significati, statistiche e metodi. *Sensus – Sensatio, VIII Coll. Intern. L.I.E. Roma 6–8 gennaio 1995, 83–119*. Olschki.

Postfazione: *Rodolfo Bozzi SJ: Introduzione al Futuro*, 2a ediz. Bari, 145–150 [second edition of the 1996 text "Rodolfo Bozzi" above].

Fare filosofia sul computer e fare filosofia con il computer. *Filosofia e Informatica – Atti del primo incontro italiano sulle applicazioni informatiche e multimediali nelle disc. filosofiche, Nov. 1995*, *Roma,* ed. Luciano Floridi, 79–84. Paravia.

1997

La terminologia come interfaccia espressivo tra pensati unici e pensanti molti. *Less. Intell. Eur. ... Atti Conv. Intern. in mem. Paul Dibon, Napoli 17–18 Maggio 1996*, 29–40. Olschki.

Il libro dei Metodi 2: Le Voci dell'Index Thomisticus raggruppate per tipi semantici e per frequenza. CAEL [pp. 402.].

Thomae Aquinatis Opera Omnia cum Hypertextibus in CD-ROM. *Cristianesimo nella Storia* 18, 397–404. Bologna.

Introduzione ai Lavori. *Umanesimo e Informatica, Atti del Convegno, Trento 24–25 Maggio*, 9–13. Ediz. Metauro

Pensieri attorno a un piccolo grande libro elettronico, *AEI* 84(9), 76–78.

1998

Il Libro dei Metodi 3: I lemmi dell'Index Thomisticus raggruppati in famiglie tematiche. Cael [pp. 437].

Dal Computer agli Angeli. 1090 Pensieri su Intelligenza Artificiale. Politecnico di Milano [handouts; 116].

Preface: *Giuseppe Prestia: Sinfonie della Vita*, 5–8. Catanzaro: V. Ursini

La Nomenclatura Musicale in Tommaso d'Aquino. *Verum et Certum – Studi di Storiografia Filosofica in onore di Ada Lamacchia*, 97–118. Levante: Bari, ediz.

Concluding a life's safari from punched cards to World Wide Web. *The Digital Demotic – Selected Papers from DRH97, Oxford Sept. 1997*, 3–11.

Oltre ai testi Tradurre anche il vocabolario di S. Tommaso. *Doctor Communis* LI(1) Genn. Apr, 80–84. Vaticano.

The translation of Aquinas' vocabulary in addition to the translation of his texts. In *Roma, magistra mundi. Itineraria culturae medievalis: Mélanges offerts au Père L.E. Boyle à l'occasion de son 75e anniversaire*, ed. Jaqueline Hamesse, 89–93. Turnhout: Brepols Publishers

Preface: *Andrea Di Maio: Il concetto di Comunicazione (...) in Tommaso d'Aquino,* Pont. Univ. Gregoriana Roma, 5–16.

La salute del cuore è la logica definitiva dell'insieme. *Il Secondo Rinascimento* 53 (Marzo), 81–86.

La storia misteriosa della salute secondo la teologia sacra. *Il Secondo Rinascimento* 57 (Luglio), 60–63.

Libro: perno su cui ruota la comunicazione degli umani e del cosmo. *Il secondo Rinascimento* 62 (Dicembre), 72–76.

1999

La voce 'signum-i' in Tommaso d'Aquino. In *Lessico Intellettuale Europeo – SIGNUM, IX Coll. Intern. Roma 8–10 Genn. 1998,* 95–117. Firenze: Oeschki.

Picture a Man ... Busa Award Lecture, Debrecen, Hungary, 6 July 1998. *Literary and Linguistic Computing*, 14(1), 5–9.

Nella pioggia di luce. *Concorso per la statua di S. Tommaso – Comune di Roccasecca,* Roma Univ. S. Tommaso 7 Giugno, p. 15.

Il Libro dei Metodi 4: L'intero lessico di Tommaso d'Aquino parcellizato in unità progressive e omogenee per elaborazioni di ermeneutica computerizzata. Cael [pp. IV + 190].

Quodlibet – Briciole del mio mulino, 169. Milano: Spirali.

Dal computer agli angeli – 1156 pensieri sull'intelligenza artificiale. 2a ediz. Politecnico di Milano [pp. VIII+120].

Il Libro dei Metodi 5: Sussidi per il Progetto del Lessico Tomistico Biculturale. Cael [pp. 183].

Perché la Chiesa nell'ammettere l'evoluzione come processo universale faccia eccezione proprio per l'origine dell'intelligenza. *The Origin of Intelligent Life in the Universe –* Proceedings of Int. Conference Varenna, 28 Sett.-1 Ott. 1998, 377–387. New Press: Como.

Tecnologie, ruscelli di sapienza. Ossia, il coraggio della certezza. *Il libro del Meeting '98: La vita non è sogno*, 26–31. Ediz. Itaca

2000

Un gesuita, l'Aquinate, Thomas Watson Sr e New York a Milano. *Logica T* 2(2) (Apr.), 17–20. Milano.

Padre Busa, la mia missione impossibile. *Tempo economico*, 38–415 (Maggio), 18–19. Milano [interview].

L'elettricità genera ancora stupore. *Il mistero genera stupore* [Meeting di Rimini 1999], 148–151. Itaca: Castel Bolognese.

Il libro dei Metodi – vol. 8.3–5: Gli archivi a disposizione del progettato Lessico Tomistico Biculturale. Cael [pp. 119].

Il libro dei Metodi – vol. 8°–2: Gli archivi a disposizione del progettato Lessico Tomistico Biculturale. Cael [pp. 205].

Dal computer agli angeli – 1261 momenti di pensiero [...]. Itaca-BVE: Castel Bolognese [pp. 253].

Presentazione del volume di A. Fagioli, Nazareno Taddei – un gesuita avanti: 17 Giugno 2000 – Bocca di Magra, 7–13. Roma: Edav.

La presenza di Dio Amore Vivente recensione di Fr. Rivetti Barbò: Dio Amore vivente – Lineamenti di teologia filosofica (Jaca Book 1997 249). *AA Vari: L'audacia della ragione – Rifless. sulla teol. filosof. di Fr. R.B*, 25–30. Roma: ed. Hortus Conclusus.

2001

Preface: *B. Bisceglia sj – A. Rizzi: Alcune analisi statistiche delle Encicliche Papali*, 4–8. Ed. Vaticana.

Fede e Scienza: 1.) il conoscere di ogni 'io' è più prendere o essere preso? o costruire? ma cosa? 2.) Ogni 'io' tra il presente, il sempre e l'evoluzione. In *AA Vari – Il Meeting 2000* vol. 1, 145–149. Itaca.

Le tre informatiche, *I nuovi media* (September), 9–14. Bologna.

Después de medio siglo el gigantesco y enciclopédico léxico de Santo Tomás continua constituyendo la vanguardia de la hermeneutica informatizada. In *XXVI Semana Tomista, Buenos Aires, September*, art. 4 [pp. 14].

Trattato di Lessicologia. Il Libro dei Metodi 6. Cael [pp. 264].

L'imprenditore e la locomotiva. In *BVE – La comunicazione bancaria, Milano – Convegno 16–17 Ottobre 2001*, NA.

Da dove vengono le idee? *Telebank – Il capitale intellettuale,* p. XLVII–XLIX. Milano: BVE.

Filosofia e Computer. *L'uomo e la macchina – 30 anni dopo – Filosofia e Informatica ieri e oggi, Atti Conv. Naz. Soc. Filosof. Ital. Bari 24–26 Ottobre 1997*, 299–307. Bari: Laterza.

Fondamenti di Informatica Linguistica parti – essenziali ancora utili, Il Libro dei Metodi 7. Cael [pp. 412 – 175 = 237].

Il computer usato bene – Intervista a Roberto Busa. *Emmeci-quadro* 13 (Dic.), 122–132 [interview].

2002

L'ermeneutica computerizzata sessant'anni dopo. Tiranë: Institutit të Logjikës dhe Etikës [in Albanian, English and Italian] [pp. 83].

Transcription of an Autograph Text of Thomas Aquinas from 1260–1265 to the present day, *Wissenschaftsgeschichte zum Anfassen*, 199–205. Stuttgart: Frommann-Holzboog.

Foreword: J.G. *Hamann Briefwechsel (1751–1788) Lexicologisches System und konkordanzen auf CD-ROM mit Thesaurus* (Roma LIE), X–XIX. Firenze: Olschki [in German and English].

Biblioteca Specifica dell'Index Thomisticus 1949 – Luglio 2002, Il Libro dei Metodi 9. Cael [pp. 181].

I due emisferi del lessico d'ogni uomo che viene al mondo: l'umano e l'interfaccia divino-umano. *Meeting 2001: Tutta la vita chiede l'eternità*, 98–104. Itaca.

Il Lessico Tomistico Biculturale, intervista a padre R. Busa a cura di Luciano D'Andrea. *Rivista* AEI 89 (Sett.), 59–62 [interview].

Experientia, experimentalis, experimentum, experior, inexperientia, inexpers: nell'Aquinate e negli altri autori censiti nell'Index Thomisticus. *Experientia – X Coll. Internaz. Roma LIE 4–6 Genn. 2001*, 101–168. Olschki.

Descrizione degli archivi elettronici esistenti su gold-disks al 31 dicembre, Il Libro dei Metodi 10. Cael [pp.162].

Hermeneutika e kompiuterizuar pas gjashtëdhjetë vjetësh, 83. Tiranë: Institutit të Logjikës dhe të Etikës [text in Albanian, Italian and English].

2003

Procedure per microanalisi di voci significanti "cose od oggetti – 1° animali". Con campione guida sul lemma A06954-apis, Il Libro dei Metodi 12. 1a edizione. Cael [pp. 67].

Quasi a modo di testamento: profezia o utopia? *Informatica e Scienze Umane – mezzo secolo di Studi e Ricerche* [Strasbourg Eur. Sc. Found. 14–15 Juin 2002], 57–72. Firenze: Olschki

La macchina di Turing mezzo secolo dopo. *La Civiltà Cattolica*, n. 3681 (1 Novembre), 238–248.

Contribution: *Il brainworker – Congr. Internaz. 30 Nov. – 2 Dic. 2001*, 116–117. Milano: Spirali.

Santo Tomas de Aquino precursor de la informàtica. In *Las fundamentos metafisicos del orden moral – XXVIII Semana Tomista – Homenaje a mons. Octavio N. Derisi.* Buenos Aires, 237–243.

2004
Il Caravaggio mai visto. *Avvenire*, 05/02, Milano, pag. 24.

De Forcellinii Lexici Totius Latinitatis ad haec nostra tempora utilitate. *Euphrosine Lisboa, XXXII in memoriam Ant. Zampolli*, 25–28.

Perché il nulla è una parola che tutti usano? Perché esiste qualcosa invece di nulla? Vuoto, Nulla, Zero. *Atti del Conv. Internaz. Villa Monastero, Varenna 30 Sett.-2 Ott. 2002*, 203–208. Ed. Itaca

Contribution: *Il cervello – Atti Congr. Intern. – Milano 29 nov.-1° dic. 2002,* 181–183. Milano: Spirali.

Review of: Adamo – Della Valle: "Neologismi quotidiani. Un dizionario a cavallo del millennio. 1998–2003, *La Civiltà Cattolica. Roma* 3699–3700 (Agosto), 319–320.

La luce umile: non fa vedere se stessa ma tutto il resto. *30 giorni* XXII(9) (Sett.), 90–92.

Foreword: Perspectives on the Digital Humanities. In *A Companion to Digital Humanities,* eds. S. Schreibman, R. Siemens, J. Unsworth, XVI–XXI. USA, UK, Australia: Blackwell.

Il Libro dei Metodi 13: Procedure per microanalisi di voci significanti cose od oggetti, 2nd ed., sez.1a: Procedure; sez.2a: campioni di elaborazione; sez.3: Prontuario. Cael [pp. 99+44+48].

2005
Il Libro dei Metodi 8(3): Gli archivi a disposizione del progetto LTB. Cael [pp. 92].

Persona, io, consapevolezza, anima: cosa sono? Qual è la risposta della Chiesa Cattolica? *Atti Conv. Naz. R.S.A.A.* Roma 2 Ottobre 2004, 62–77 [with a version in English].

Il Libro dei Metodi vol. 14: Stralci di prova della Concordanza in ordine di testi del lessico economico di Tommaso d'Aquino, di prossima pubblicazione, Cael [pp. 87].

Preface, *Giuseppe Prestia: I Canti del Romito*, 7–13. S. Maria di Catanzaro: ed. Ursini.

Universo e Multiverso, *Dialogu Sot* (Giugno), 24–30. Tirana. Ist. A. Zuli.

Omnes latinasque explicationes ex 56 voluminibus Indicis Thomistici [...] curaverat Robertus Busa [...] Libro dei Metodi 16. Cael [pp. 325].

Machina, mechanicus, machinatio, machinor nell'Index Thomisticus. In *Atti XII Coll. Intern. LIE, 8–10 Gennaio 2004*, 171–239. Firenze: ed. Olschki.

2006
Lingue Disciplinate potranno rispondere alla sfida linguistica della globalizzazione? Libro dei Metodi 17. 2a ediz. Cael [pp. 96].

Procedure per [...] cose od oggetti. 2a ed. sez. 4 Libro dei Metodi 13(4) "Mansionario". Cael [pp. 11].

Thomae Aquinatis De Re Oeconomica contextus omnes ordine textuum [...] ordinavit R. Busa. Napoli: Istituto Italiano Studi Filosofici [3 vols., compl. XXIV+1692; Tomus I – praefationes, indices, appendices, X+402].

Tomus II – contextus ex operibus propriis [pp. VI+758].

Tomus III – contextus ex comment, reportat, dubiis [pp.VIII+532].

Intervista a padre Busa. *Gesuiti d'Italia* 5 (Sett.–Ott), 402–403 [interview].

Review of: Sergio Galimberti: Aurelio Andreoli e il Centro Cultrale Veritas di Trieste, 492–496. MGS Press: Trieste.

Rovesciando Babele – ossia tornare alle radici di ogni lingua. Milano: Spirali [pp. 228].

Lingue disciplinate – potranno rispondere alla sfida linguistica della globalizzazione? In *Atti della Accademia Nazionale dei Lincei, classe di scienze morali, storiche e filologiche, anno CDIII*, pp. 385–399. Roma.

Differenza e Diversità, *Enciclopedia Filosofica* (Fondazione Centro Studi Filosofici di Gallarate), vol. 3: 2.842–2.844. Bompiani.

- Distin zione, ivi, 3.003–3.014, vol. 3
- Divisione, ivi, 3.029–3.334, vol. 3
- Lippert Peter, ivi, p. 6.518, vol. 7

Contribution: A noi piace solo il bello. *Art Ambassador – Forum mondiale di Informatica* 1(4) (Dicembre), 252–255. Milano: edizioni Spirali.

Lingue disciplinate potranno rispondere alla sfida linguistica della globalizzazione? Libro dei Metodi 19, 3a edizione, Cael [pp. 29.].

2007
Contribution: Tornare alle radici di ogni lingua. *Il valore dell'Italia – Festival della Modernità*, 30 Novembre – 3 Dicembre 2006, 184–186. Milano: edizioni Spirali.

La microanalisi sintattica va condotta su 2 livelli – Parte Prima, Libro dei Metodi 20, Cael [pp. 65.].

La dottrina delle quattro cause a ricordo di Naz. Taddei sj. *EDAV, La Spezia* 351 (giugno), 9–10.

Postface: Informatique et humanité numériques. Claire Brossaud & Henry Bakis, *Humanités Numériques* 1, 272–279.

Postface: Informatique et humanité numériques. *Humanités Numériques 2: socio-informatique et démocratie cognitive*, 203–205. Paris: Lavoisier

Contribution: *La Scrittura – Milano 24–27 Maggio 2007*, 282–283 e 543 [interview].

2008
Contribution: A. Verdiglione et al.: *Il Cervello e la Bussola*, 15–16. Milano: Spirali [interview].

Scienze linguistiche e informatica. Rovesciando Babele. Incontro con padre Roberto Busa. *Green: La Scienza al servizio dell'Uomo e dell'Ambiente* 12 (Marzo), 4–9.

L'incompiuto mio ultimo Quodlibet – vol. 1, Libro dei Metodi 21(1) [pp. 66].

L'incompiuto mio ultimo Quodlibet – vol. 2, Libro dei Metodi 21(2) [pp. 261].

L'incompiuto mio ultimo Quodlibet – vol. 3, Libro dei Metodi 21(3) [pp. 486].

L'incompiuto mio ultimo Quodlibet – vol. 4, Libro dei Metodi 21(4) [pp. 554].

Metafora. Preface: A. Bandiralli, *Genesis – I giorni della creazione*, 7 [private edition].

Memorie Informatiche di un Ottuagenario. *Moderni e Antichi – Studi sul Classicismo*, 445–451. Arezzo.

Intervista di F. Rizzi a padre Busa su Papa Luciani. *Avvenire* 27(08), p. 21 [interview].

La libertà delegata. In *La libertà*, 275–276. Spirali.

D. Bamman, M. Passarotti, R. Busa, G. Crane. The annotation guidelines of the Latin Dependency Treebank and Index Thomisticus Treebank. The treatment of some specific syntactic constructions in Latini. In *Proceedings of the Sixth International Conference on Language Resources and Evaluation (LREC 2008)*, 71–76. Marrakech, Morocco: ELRA.

2009
Preface: S. Sangalli, *L'analisi dei vizi capitali in San Tommaso d'Aquino*, 124; *L'Osservatore Romano* 16–17 (febbraio), p.5. Libreria vaticana.

Chapter 20 "A Tall, Stooping Figure in Black Crossing the Courtyard": Philip Barras' Recollections of Roberto Busa S.J.

Philip Barras and Julianne Nyhan

Editors' note:

This oral history interview, between Philip Barras and Julianne Nyhan, was held on 2 March 2017 in Foster Court, UCL (in Julianne Nyhan's office). The interview was recorded, and subsequently transcribed by Jessica Salmon. Julianne Nyhan and Philip Barras then iteratively edited and revised the transcript, so as to make its style and form closer to written English.

In this interview Barras recalls his first encounter with Busa and his experiences of working with him and translating many of his publications. Barras' discussion of how he went about translating Busa's work, and of the hallmarks of Busa's writing style, offers a rich insight in to the translation process. Busa and Barras became close friends; in the course of this interview Barras recalls some charming stories about Busa, like his skill in lock-picking.

References

Busa, R. 1994. *Inquisitiones Lexicologicae in Indicem Thomisticum.* A Roberto Busa S.I. latino sermone confectae atque a Philip Barras in anglicum sermonem translatae. 2a ed. emendata auctaque. Gallarate: CAEL.

Julianne Nyhan: How did you first encounter Busa?

Philp Barras: He had written an account in Latin, *Inquisitiones Lexicologicae in Indicem Thomisticum* (Busa 1994), of how he compiled his computerized lexicon of the works of St Thomas Aquinas, and he wanted it translated into English and published in parallel text. He asked at the Università Cattolica del Sacro Cuore, Milan, where I was a *Lettore* teaching English, if anyone could translate it into English, and they suggested me.

I'd never met a Jesuit before and was rather dreading the prospect of doing so. He came around to where I lived one gloomy afternoon in October. I remember looking at a tall, stooping figure in black, crossing the courtyard, and my heart began to sink. To my surprise we hit it off right from the first, and I made one of the best friends I've ever had.

JN: You started working with Busa in about 1992?

PB: I worked with him often between 1991 and 2003, and then I did several things for him after I came back to England. Whenever I went back to Milan, (I go 3 or 4 times a year to see my daughter, who lives in Milan), I always used to make a point of going to see Padre Busa at Gallarate, until he died in 2011, and even now I still go back to Gallarate to put flowers on his grave.

J. Nyhan and M. Passarotti (eds.), *One Origin of Digital Humanities*,
https://doi.org/10.1007/978-3-030-18313-4_19

JN: Tell me about the time he brought you to the Aloisianum in Gallarate (the place where Busa lived and worked for much of his life) and you stayed there for 3 days and worked with him.

PB: The Aloisianum is a lovely place. All the Jesuits there were friendly, and they made me very welcome at meals, which were simple, good and wholesome. Of course, I was working in the afternoon, so I had to give the wine a miss at lunch time, for I needed all my wits about me! The wine, by the way, was very good indeed; it was a Barbera from Pavia and sent over gratis. And that's where, later on, I met Cardinal Martini, who'd retired there. One day Padre Busa said "Look, let's have lunch with Cardinal Martini", whom he'd taught. That was a lunch to remember. Cardinal Martini insisted on speaking English throughout, and perfect English as well.

JN: Busa spoke English well?

PB: Certainly, although he said his English was "cowboy English" because he'd spent more time in America than in England.

JN: Which languages did Busa speak that you know of?

PB: Fluent English, fluent German, fluent French. I went to see him one day with my daughter, who was about 2 or 3, and our babysitter, who was Peruvian, came along too. As soon as Padre Busa saw her, he went straight into fluent Spanish, and I thought "there's no justice in this world!". He probably spoke Portuguese as well, for two of his articles in this book was written in it (see Chaps. 13 & 16). Beyond that I can't say, but it wouldn't surprise me if he spoke half a dozen others as well. Incidentally, one of his favourite authors for bed-time reading was P. G. Wodehouse in English. When you think of how idiomatic and mannered Wodehouse is, the ability to appreciate that argues remarkable linguistic skill.

JN: I gather that the material you translated for Busa was quite difficult?

PB: It was indeed. That was why I had to sit with him, so that he could explain the concepts. He explained that he did this because he'd once had one of his articles translated, but hadn't checked the translation before it was published, and his colleagues ribbed him about it. So, ever after that he insisted that the translator sat with him so that he could explain everything, and quite right too. Once I'd been working for him for a few years, I could do his Italian on my own. But the Latin in that book! Well, I could see how all the sentences, all the syntax, hung together, but even after I'd put it into passable English, I couldn't understand what it was about. So, he patiently took me through it and explained all the concepts. I couldn't possibly have translated all that on my own, without him. At the end, he said, "do you mind if I put your name on the front?" I said, "I'd rather you didn't, because I've really been your amanuensis". He was the one who translated it, not I. It was very generous of him to put my name on the front.

JN: I'm sure you were very generous with your time as well. Aside from the person you've already mentioned, did you encounter or hear him speak about translators with whom he had previously worked?

PB: No.

JN: I ask because, having spent some time over the last weeks looking through some of his papers, I don't think he often included the name of the translator on the translated work. This makes it quite hard to figure out the identity of the other translators with whom he worked. I haven't seen any mentions of them in the material I've looked at so far, but I'm pretty sure that he must have been working with translators from an early stage?

PB: He must have done. Although the fact that he enquired if anyone could translate his book from Latin into English suggests he didn't know any English speakers who could translate from Latin. He must have had translations of his articles made, because his work was pioneering and so important. Not every computer expert, I should imagine, can understand Italian.

JN: I've also noticed that he would publish an article in Italian and then it would sometimes be re-published in English and German, sometimes in French as well. Looking through his bibliography, his publications were translated into, or written in at least 5 languages, if not more.

PB: Which are those languages? English, German, French …

JN: Portuguese and Hebrew too; Latin, of course, so that's actually more than 5. So, I'm assuming he can't have done all those translations himself?

PB: Probably not, although it wouldn't surprise me if he had, for his linguistic skills were so great.

JN: And you mentioned that the two of you were working from 8am to 8pm?

PB: That was just for 3 days in 2005 in the Aloisianum. But quite often in the 1990s, if I had the time, I'd spend the whole day with him.

JN: Did you meet any of the people who had worked with him as collaborators or in any other capacity? Or his secretary?

PB: Only latterly, a few years before he died. Sometimes at Gallarate people who'd obviously worked for him would come and sit with us at lunch. But I can't remember what was said.

JN: So, what was the experience of translating for him like? I think you've already mentioned that you found it to be an enriching experience?

PB: Absolutely, I learnt so much. It was marvellous, we'd talk about all sorts of things. Apart from the fact that I was in the company of a wonderful human being, it was a splendid intellectual challenge. He kept me on my toes and made me think. It was rather like sitting with my old tutor at college.

JN: And did he tell you about the significance of his work?

PB: Yes, he did. I'm not sure I understood it. I'm a bit thick when it comes to computers!

JN: Is it your recollection that he explained the significance of his work to you in computing terms?

PB: No, he explained it very, very simply. He obviously knew that I didn't have the technical vocabulary. He had a gift of explaining things—going to the essentials and explaining them simply. You would spend an hour with him and you would come away feeling that you understood everything. Of course, you didn't, but he put it so simply that anybody could grasp the concept at least.

JN: How I wish I had that skill! How did Busa respond to the translations that you made for him?

PB: He was fascinated by some of the English that I used, and so I'd like to think he learnt a bit about English from me. Let me explain: I would sometimes use the subjunctive, which is pretty uncommon these days, a little elegant touch, and he really liked that sort of thing. In one of the translations that I've just done [for this book], he has a metaphor: "I imagine myself as a little tugboat taking out a great ship from harbour … into the vast ocean" (see chap 19). In my translation I made sure that when I was talking about the tugboat I used words of one syllable that were all of Germanic origin, and then when the great ship goes out onto the ocean, I used polysyllabic Latinate words. So, we were always talking about little things like that. Well, that's been my trade for the past forty-odd years.

JN: Regarding the translations that you recently did for our book - you didn't have Busa with you, so how was that experience for you?

PB: It was a little bit uncanny because it was like having him in the next room, and I could almost hear him saying what he would have said, and imagine what posture he would have adopted. When he was thinking, he would always put his hands together and look up to heaven, as if for inspiration. He would be working things out in his head, and then say, "Right, 1, 2, 3, 4, 5…", and it would come out so logically, so beautifully simple. Fortunately, when I ran into difficulties, I could ask for help from a former Italian student of mine, who is now a lecturer at Nottingham Trent University. Padre Busa's Italian is beautiful, but it's sometimes too elegant for my understanding. I could also ask my ex-wife, who's Italian – she's now a professor at the Catholic University in Milan. And sometimes neither she nor my former student could really understand him. We managed to hammer something out, which is why I'm grateful you and the others are reading through the translations I've made and checking that I haven't made any howlers about technical terms with computerized linguistics.

JN: I have to say that we sometimes find it very difficult too and no, I don't think you've made any. Sometimes there's no "right" or easy answers because the translations will actually have shifted in the time since a given article was written, which is an added layer of complexity …

I often wish I could read the Italian because of the sense of distance that I feel from the original texts. But your translations, I think, really communicate the elegance of his writings.

PB: Thank you. I've tried to be as clear as he would have been, or at least as he explained things to me, and I've tried to be elegant as well. So, what you have is my prose but with his input.

JN: About the times when even native speakers of Italian found the passages difficult to interpret, is that due to the wordplay, the concepts, or maybe the different layers of language and concepts?

PB: All of those. Italian can be tortuous; some Italian academics have little to say but take ages to say it! He had a lot to say and said it extremely well, so his prose, as far as I can judge, is beautiful. But yes, the concepts were so difficult. I've seen a lot of Italian academic writing, and I had translated a fair bit of it before I started working for Padre Busa. Some of the articles and books I had translated were abysmal drivel and should never have been written in the first place, let alone published. But I suppose you could say that about some academic writing in any country. Say I used to get, not from him, but from others, 20 pages of Italian, and I gave back 12 of English. They would say, "where's the rest?" and they wouldn't believe me when I said, "well, English is far more concise than Italian". I tried to express things simply, and they couldn't understand that this is how you do things in English. Italians generally never use one word where fifty will do. Let me give you an example. When I first went to Italy, to Milan in 1987, I taught English in a private school. I was reading Shakespeare's *Julius Caesar* with the sixth form, and one day I said, "For homework write me a little biography of Shakespeare." One girl produced the following sentence: "He followed the course of education relative to the grammar school of Stratford-upon-Avon." She was horrified when I crossed this out and wrote, "He went to the grammar school in Stratford-Upon-Avon". She said, "You can't say that!" She was obviously thinking in Italian, and that's how it would be expressed in Italian. If in Italian you say merely, "He went to the grammar school", that's not elegant. So that is why I say that Italian can be somewhat prolix.

JN: But in Busa's writings, by way of contrast, the concepts may have been very complex, but the way he expressed them was …?

PB: He didn't waste words; every word told and was in its proper place. After one day with him going through his Latin, which was golden classical, I thought, "I'd give anything to be able to write Latin like that!" On another occasion, he lost the thread and put a singular verb where there should have been a plural. I pointed this out, and he wasn't pleased with himself. When I went home that evening, I said with a touch of *Schadenfreude*, "How marvellous! I spotted a mistake that Padre Busa made in his Latin!"

JN: And then you could jokingly call him … what was it he called you? *Professore*?

PB: Yes, at first, particularly when I was being a bit thick. Then later on, after a few years, he said, "Would you please do me a favour and use "*tu*"?", the friendly second singular rather than the formal third singular. I said, "thank you, but I can't. First of all, I'm 40 years younger than you are. And secondly there's the reverence I have for your priestly orders." He just looked and said "try to get used to it". After that, well, I always addressed him as "*Padre*", of course, but always used "*tu*" …, which was a great honour.

JN: Would you tell me a little more about Busa's personality?

PB: He was just so kind; very witty in a gentle sort of way. People told me that when he was young they thought he was a bit severe, but he'd mellowed an awful lot. After I got divorced, and then married again in England seven years later, he said, "well, if I can make it, I'm looking forward to coming to your wedding." The funny thing was, apparently, he would never have said that forty years before. Of course, he knew everything about human nature, there can't have been much that he hadn't seen. I once went to a conference in Milan with him which he had organized. I think it was when the first edition of *Inquisitiones Lexicologicae* had come out, and such people as Umberto Eco came (Padre Busa invited me to have lunch with him and Umberto Eco, who actually was a very gruff man whom I didn't like very much). When Padre Busa was sitting at the front of this conference, he said something like, "well yes, I've listened to all your praises of my work, but that's only secondary: never forget that first and foremost I am a priest in Christ's service, and that means 24 hours a day, 7 days a week".

JN: This was something else I also wanted to ask you. I think it's unavoidable in his writings, isn't it, the sense of higher purpose?

PB: Absolutely. Well, you must have been through all your doubts by the time you become a Jesuit. You need absolute certainty to say, "the Lord is always talking to me, he will see me through." He was always quoting the Bible; it wasn't laboured, it was just natural, that's how he was. And he always looked for the good in everybody. In 2003, because of marital difficulties, I'd left Milan and returned to England. I hadn't seen him for six months, but wrote to him to tell him that I was coming back to Milan for a few days. He was about to go to Rome and he said "come up and see me". The only time he had was early one Sunday morning before he was due to take a train to Rome at half past ten, and so he said, "can you come and see me at half past eight." He gave me over an hour on that Sunday morning, and said, "don't despair, I'll try to fix things."

With him, always when I was in his company, I felt myself to be good. Not as in the common phrase "I felt good"; I felt myself to be a good man, because in his company you could be none other, he just inspired you. When I was with him, there was no doubt in my mind what was right in any given situation. When I'm on my own, I think, "what would he do?", and that inspires me. To quote the

Psalms—I think it's one of the psalms—"*Lucerna pedibus meis*" [Psalm 118/119: 105], a light for my feet.

JN: I think I've asked you most of the questions I wanted to ask...

PB: Let me tell you another couple of stories. When our daughter was to be christened, my wife went to the local parish priest and asked if this could be done when my parents were over in September 1994. She was turned away with a harsh answer. First of all, they didn't do single baptisms, it was always in groups of twenty. Secondly, since I was an Anglican, I would have to undergo six months of Catholic instruction. When she came back and told me this one Friday morning, I was, the Italians have a charming saying: "so angry that I was beyond the grace of God". Fortunately, I was going to see Padre Busa that afternoon; when I told him, he smiled and said, "leave it with me." The next morning he rang me up and said, "right, I have two dates in September at the Jesuit church, San Fedele", (which is just down from the Galleria, from La Scala; the façade is splendid; it was there that the first performance of J. C. Bach's Magnificat and Mass took place in 1760; he was organist at Milan cathedral before he came to London), "which do you want?" I said, "which date would you prefer, Father?" And he said, "well, at my age, if there's a choice, I always take the first!" A very gentle wit. He did the most of the service in Latin, with some prayers from the Book of Common Prayer at the end.

On a later occasion, at Gallarate, just a few years before he died, I'd gone to the jakes after lunch and when I tried to get out, the lock jammed. I hammered on the door shouting for help, and he came with paper knife and expertly picked the lock. I said to him, "If you hadn't been a Jesuit, you could have had a lucrative career as a burglar." To which he replied, "You never know what you might have to do in my line of business."

JN: And he was an active priest, an active Jesuit the whole time he was pursuing his scholarship and managing the centre and raising funds and ...

PB: Yes. Every day he must have been on his feet, or knees, from six in the morning until ten at night. And, apart from his studies, he always had time to see people. I don't think he ever rested during the day until his last few years, when he was ordered to do so.

Index

J. Nyhan and M. Passarotti (eds.), *One Origin of Digital Humanities*,
https://doi.org/10.1007/978-3-030-18313-4

M

N

P

R

S

T

U

V

MIX
Papier aus verantwortungsvollen Quellen
Paper from responsible sources
FSC® C105338
FSC
www.fsc.org

If you have any concerns about our products,
you can contact us on
ProductSafety@springernature.com

In case Publisher is established outside the EU,
the EU authorized representative is:
Springer Nature Customer Service Center GmbH
Europaplatz 3, 69115 Heidelberg, Germany

Printed by Libri Plureos GmbH
in Hamburg, Germany